SACRED NUMBER AND THE LORDS OF TIME

The Stone Age Invention of Science and Religion

RICHARD HEATH

Inner Traditions
Rochester, Vermont • Toronto, Canada

Inner Traditions
One Park Street
Rochester, Vermont 05767
www.InnerTraditions.com

Library of Congress Cataloging-in-Publication Data

Heath, Richard, 1952–
  Sacred number and the lords of time : the stone age invention of science and
religion / Richard Heath.
      pages cm
  Summary: "An exploration of sacred geometry, space, and time encoded in stone
structures during four successive ages of megalithic building." — Provided by
publisher.
  Includes bibliographical references and index.
  ISBN 978-1-62055-244-5 (pbk.) — ISBN 978-1-62055-245-2 (e-book)
  1. Science, Ancient. 2. Religion, Prehistoric. 3. Astronomy, Ancient. 4.
Megalithic monuments. 5. Carnac (France) 6. Stonehenge (England) 7. Pyramids
of Giza (Egypt) 8. Teotihuacán Site (San Juan Teotihuacán, Mexico) I. Title.
  Q124.95.H43 2014
  510.9'013—dc23
                                                                    2013037814

Printed and bound in the United States

10 9 8 7 6 5 4 3

Text design by Virginia Scott Bowman and layout by Priscilla Baker
This book was typeset in Garamond Premier Pro with Copperplate and Gill Sans
used as display typefaces

To send correspondence to the author of this book, mail a first-class letter to the
author c/o Inner Traditions • Bear & Company, One Park Street, Rochester, VT
05767, and we will forward the communication, or contact the author directly at
unigram@btconnect.com.

*In memory of Alexander Thom, Hélène Fleury, and
John Michell, without whom this book
would not have been possible*

CONTENTS

ACKNOWLEDGMENTS

THE READER MIGHT wonder why the story told in this book has not
been possible until now. The reason is the dispersed nature of essential
aspects of this story, for no story can emerge without a reasonable explanation of how one thing could have led to the next. In this way one
forms a possible history of the megalithic.

The fundamental steps that led to making this story were the discovery of day-inch counting with my brother Robin Heath and the
subsequent discovery of the importance and uses of circumpolar observatories in the megalithic. The work of John Michell and John Neal in
pursuit of ancient metrology has been invaluable. Since the 1970s they
have been restoring historical metrology into a pre-arithmetic numerical
science. Alexander Marshak's analysis of time-factored bones has been
invaluable in demonstrating that counting time was a widespread hobby
if not a tool in the Upper Paleolithic period. The Association Archeologique Kergal (AAK), a group operating in the 1980s to research in
particular the monuments of Brittany near Carnac, contributed through
their magazines unique reportage on monuments with useable technical
details and demonstration of the use of multiple squares. Howard Crowhurst, a former member, first showed me monuments in 2004, and he
has built on their work with tours, conferences, and books. AAK draw-
ings have been used throughout chapter three. The engraved stones of
Carnac were scanned in 3D by Laurent Lescom as part of a Nante University team, and two of these were useful for adding an extra perspective on day-inch counting and resolving the royal cubit's astronomical

origins. One has to thank the authors of *Hamlet's Mill* for their efforts investigating the mythology of precession, integrating the measurement of the earth with the astronomical studies of the megalithic period. The downloadable Google Earth database from megalithic.co.uk provided a comprehensive guide to geodetic menhirs not available in books. John Michell's correspondence and work on the quarter-degree model from Stonehenge to Avebury was key to chapter six. John Neal's analysis of the Great Pyramid and geodetic works in Egypt gave a direct clue to megalithic transmission of metrology to ancient Egypt.

INTRODUCTION

MODERN MATHEMATICAL SCIENCE deals in precise measurements accurate to many decimal places. Simple integers rarely appear. The trend has recently been toward reforming our units of measure to get away from specific objects of reference and base them on universal physical properties. In ancient times people tried much the same thing, but, not having an arithmetical system, they used whole numbers of the same length (the inch) to measure astronomical time (the day). Then, using geometry, they created their first objective measure, a megalithic yard, which expressed the difference between the solar and lunar year. Their idea of sticking to whole numbers remains part of our number theory and, as Leopold Kronecker famously said, "God created the natural numbers, all else is the work of man."

The natural numbers or integers carry with them a sense of unity and design as to how they interact with one another. As *symbols* these number relationships affect the physical world and this suggests they provided a fundamental creative fabric for the universe. The constructions made by megalithic people present such a view. The monuments could only reflect a "heavenly pattern" ("As above, so below") because the fabric of abstract whole number relationships appears to have been employed in a later weaving of planetary time cycles, which were then seen as the work of some god or gods (the demiurge) who surrounded the Earth with numerical time ratios.

Today it is not easy to believe in an organizing role in the world between whole numbers and the structure of the physical systems that surround us, the Earth, and the solar system. However, many of our

1

institutions are relics of a time when human life was organized to correspond with such a reality, as with the historically widespread seven-day week, the twelvefold division of space, the ideas behind many of our religious symbols, the organization of ancient and modern calendars, and the units of measure we still use. Those who investigate this phenomenon of *ancient number association* must either assume these were merely ancient beliefs without a substantial basis in fact or be damned (as a regressive influence within a modern scientific age) for suggesting that the world has a numerical design.

However, such polarizations of thought usually hide great riches, despite being sat on by the proverbial dragons of cultural prejudice. The ancient number theory was a system of thought at least as grand as ours that connected human thought and action to the causes of the universe itself, and did so in a more direct way than our own science or religions do for common citizens today. Indeed, this ancient science appears to have invented our religions and developed the numeracy underlying our modern science.

The number relationships of the following chapters imply one of two things. Either the planet we call Earth has been created according to some numerical criteria or some incredible genius in the past invented an amazing numerical discipline to make it look as though the Earth had been created by inventing a system of measures, a map of astronomical time, and a model of the Earth's geoid (its shape and size) that conform elegantly to several numerical relationships. This ancient geoid was remarkably simple, being based upon different approximations to pi and made implicit within ancient metrology through the cunning use of whole number ratios. One has to question whether it was possible to have created such a wonderful artifact without the Earth actually being like that.

It might have been fitting for early man's intuitive powers to have been awakened in this way, as an artifact of creation. It appears unlikely that the massive bodies surrounding the Earth could have accidentally fallen into time cycles so rationally interrelated, especially when the earth is not at the center of the solar system. It stands out that the entire solar system is somewhat focused on the Earth, making its cycles fit some meaningful criteria just as the shape of the earth is found related to time.

The factual basis for creationism is not at odds with modern science but rather with a modern superimposition, which mistakenly concludes that the gods of religion did not, in any way, make the world (just as we alter it using physical forces). But eliminating something from possible proof is fundamentally unscientific and, in this case, an unfortunate reaction against an ancient world where the knowledge of such gods was an exact science like our own.

If some ancient group or groups managed to read astronomical time and the shape of the earth, then their genius was already old by the time of the ancient Near East, whose numerical sciences were derivative but now definitive, defining what any ancient astronomy could have been. I suggest that what followed the original astronomical discoveries, recorded in the megaliths, was a descent into superstition about the gods and an ascent into today's mathematics, focused on the calculations and proofs possible once numbers were liberated from being the lengths within geometrical structures.

The scientists of the megalithic can then be seen as having had a distinct purpose for the evolution of the human intellect, to establish both an early exact science and to develop our religious symbolic cultures, where symbols were religious intermediaries, expressing the form of the gods. The megalithic sciences may also have been an initiation for the human mind, anticipating the kind of cultural mind we have today and this, like many real-world initiatives, probably met with complications due to the behavior of the human groups aggregating around the new technologies of, first, a Bronze Age and then an Iron Age.

Now relegated to our cultural unconscious, we need the megalithic culture to be better appreciated today, as they give a deeper explanation for why this planet and its living systems are special within the universe. Technology appears to be "dumbing down" the post-industrial population into mere users and consumers of convenience and entertainment, unconnected to the cosmos. It further appears impossible to politically resolve the set of impacts we are inflicting upon the natural environment, potentially marring the living planet. Religious beliefs are often so dissociated from any megalithic origins that faith is often lost or dogmatic. Most importantly, the functional model of human action, compatible with our scientific experimental methods, is commonly believed to have disproved any cosmic purpose for a human life.

The megalithic discovery of the world as a purposeful creation within the cosmos was also a discovery that human development, like that of the Earth and the planets, needs to follow a law-conformable path. The lost paradigm of number association, organizing the world in an intelligible way, has not been detectable to modern astronomy, and, indeed, the peer review process rejects any theory that lacks a physical causation. The whole point about ancient number science is that it looked at numbers as organizing principles rather than as the outcomes of physical measurements.

In my research I've found that understanding megalithic monuments conveys a sense of theater, for such monuments were often staged productions rather than operational structures. The builders of these monuments took care to enhance their pedagogical veracity, even for unknown future audiences such as visitors from our own time. The theatricality becomes clearer when the exact techniques required to encode the depth of information into the monuments are properly recognized. The large stones we associate with the megalithic period would only have gotten in the way of actual megalithic activities, which instead used tools consisting of bone, wood, stone, and fiber. The stone monuments were largely retrospectives upon their scientific work and only represented their implicit methods of working.

We are faced today with a sad fact, that the megalithic has been ill appreciated by recent centuries because of our own competing interests and attitudes emerging from religion and science. Unlike ourselves, the peoples of the megalithic did science *before* religion and almost certainly created their mythologies through new types of exact knowledge of sky and earth. Their developed language, visible at megalithic sites as alignments, geometry, and exact measures, has a contemporary role in leaving open a last chance to see the foundations of our own civilization. Some of these large stone buildings still exist solely because of the use of megaliths, as if to preserve their knowledge despite coming changes.

Examining these megalithic sites, one can glean perspectives about these enigmatic precursors of civilization and the megalithic mode of human existence. Integrating their new knowledge into humanity's oldest stories, we can discover not only megalithic ideas about the role and status of human beings in the universe but also the essence of the megalithic imagination that preserved these ideas in stone for millennia. I propose that modern cultural life would benefit at this time from a now-freshened renaissance of the prehistoric worldview.

PART ONE

A Journey to the Lords of Time

The Ishango Bone, a bone tool dated to the Upper
Paleolithic era and now believed to be more than 20,000
years old. It has been interpreted as a tally stick but also as
a medium for a Stone Age awareness of prime numbers. It
was found in 1960 by Jean de Heinzelin de Braucourt while
exploring what was then the Belgian Congo.

1

THE AWAKENING OF
THE STONE AGE

THE STONE AGE in Europe was an extremely stable mode of existence for Cro-Magnon humans, lasting from 40,000 years ago until around 5000 BCE, just two thousand years before our present civilized history began. These last two thousand years of "prehistory" left some remarkable megalithic monuments, especially in northwestern Europe. The megalithic period in northwestern Europe was contemporaneous with the Neolithic period in the Near East, which preceded the ancient civilizations of the third and second millennia BCE. Thousands of miles northwest of the Neolithic Near East, the European megalithic period shared the benefits of a climatic optimum encouraging a more settled way of life. While climate, patterns of settlement, and other influences transformed the European Stone Age into the first megalithic age, further south agrarian surpluses became possible and these would power the city-states and theocracies of the ancient world.

It is probable that a large population of workers could be supported at more southerly latitudes, conditions also true when central and southern Mexico had their pyramid-building periods in the first millennium CE. The megalithic era of northwestern Europe was probably less settled and didn't develop to resemble the historic Near-Eastern civilizations of the Egyptians and Babylonians. However, the pyramids and ziggurats of the Near East employed similar units of length as well as the same astronomical and geodetic knowledge first deployed thousands of years earlier within the megalithic constructions of northwestern Europe. A channel of transmission must have existed that allowed identical units of measure to travel, units like the Egyptian royal yard. This "royal" unit of measure is found inside the Gavrinis chambered cairn (France, 3500 BCE) as an engraved royal cubit, in the Sarsen Circle of Stonehenge (England, 2500 BCE) as the royal yard, and at Teotihuacan (Mexico Valley, after 100 BCE). The length of this unit at these locations is so exact as to rule out its independent generation, especially because the royal cubit is itself based upon a unit established in megalithic Europe and crucial to the design of the metrological system as a whole: the present-day foot.

Thus it is that the pyramid-building cultures in particular must have been strongly influenced by the discoveries of the megalithic period of northwestern Europe between 5000 and 2500 BCE. Mexico's pyramids came two millennia after those of the Near East, which in turn came two millennia after the early megaliths of northwestern Europe. It appears that for the last seven millennia, the metrological system used for all civilized building works has been that first developed by the European megalithic between 5000 BCE and 3500 BCE. This megalithic culture somehow transmitted its metrology to the later civilizations, starting with the ancient Near East sometime between 3200 BCE and at least one hundred years before the Great Pyramid of Giza demonstrated its megalithic metrology (2560 BCE). The monuments of northwestern Europe show that metrology took at least two thousand years to develop. This transmission of metrology to the ancient Near East has needed an explanation for many years but has miraculously eluded the contemporary account of history.

In a reversed diffusionism therefore, from West to East rather than East to West, the builders of the Great Pyramid of Giza used units of measure already developed in Brittany and Britain, such as the royal foot, cubit, and yard. The royal cubit measure of 12/7 feet is to be found engraved at the end chamber of Gavrinis in southern Brittany, a monument built around 3500 BCE and intentionally sealed around 3000 BCE. But this intrigue concerning our metrology goes deeper than its transmission into the Near East from the early European megalithic period. The megalithic metrological units, such as the royal cubit found in Egypt, have an exact fractional relationship to the English foot as well as to features of the Earth's dimensions, such as its equator, meridian length, and polar radius. This reveals an ancient and intentional derivation of the English foot from the dimensions of the Earth.

For example, the length of the Earth's equator is 131,487,192 feet—that is, 360,000 times 365.2422 feet—so the English foot appears to have been defined according to the length of the Earth's equator, 360 degrees in a circle, and a primary astronomical period, the 365.2422 days in a solar year. This attempt to find such a concrete unit of measure relating to the Earth itself is similar to the derivation of the megalithic yard, which relates to the solar and lunar years when counted in days (see chapter 3).

The actual length of the equator was divided into a number of parts equal to the solar year (365.2422 days), and each solar-year-part was divided into day-parts* with a distance of 360,000 feet for each day (or 72 miles of 5,000 feet). Such measurements were possible during the megalithic period, the technologies clearly demonstrated in Carnac's monuments, built in the fifth millennium BCE. This established relationship of the foot to the equator enabled the metrological system to operate in a very creative way between the whole numbers 360, 365, and the real number of days in the solar year, 365.2422. There is evidence in surviving traditions for the division of the equator into 365¼ parts. As John Michell states in *Ancient Metrology:*

> According to Needham,† this method of dividing a circle into 365 ¼ degrees was applied in ancient China to the circle of the equator. In that case, taking the length of the equator to be 131,490,000 ft. or 24903.4 miles, each of the 365 ¼ degrees of equatorial longitude contains 360,000 English feet, each minute 6,000 ft. and each second 100 ft.[1]

---

*The day-part refers to the angle through which the sun moves in a single day, requiring the Earth to rotate a small fraction extra over a sidereal day, this adding about four minutes to arrive at the normal solar day.

†One presumes Michell is referring to Joseph Needham in *Science and Civilisation in China: Volume 3, Mathematics and the Sciences of the Heavens and the Earth*, Cambridge University Press, 1959.

Evidence of such a geodetic basis for metrology can be seen in the key dimensions of the Great Pyramid (2560 BCE), which form a scaled model of the northern hemisphere of the Earth. Its height represents the polar radius—the length of which could only have been inferred with some difficulty using a geodetic survey between two latitudes, one in the equatorial region south of Egypt. John Michell found that the British megalithic monuments of Stonehenge and Avebury had been placed one-quarter of a degree in latitude apart on a north-south line so as to form a non-vertical model of the Earth's three key radii: the equatorial, polar, and mean radius (see chapter 6). This would have been possible without a full survey, but only at that special latitude.

Such a geodetic work was only achievable through the use of a metrological system in which many units of measure divide into the Earth's dimensions in interesting ways, such as how the root foot was defined through division of the equator's circumference by the number of days in a solar year and the 360 degrees of a circle. The reason for this metrological system's success has become a reason for enforcing its obscurity in our scientific age, for metrology appears to have been designed with a prior knowledge. The ability to measure the Earth in an ideal and intelligible way was designed before the Earth was actually measured. It is as if the Earth had itself been "designed" and that the metrological system was a part of the Earth's design. A simpler and more likely explanation is that metrology encountered its problems in a near perfect order, enabling numerical decisions to be made that would prove numerically ideal as each further dimension of the Earth became known.

The geodetic work from England arrived in Egypt before 3000 BCE. Over one thousand years before this, in Brittany, astronomical time had been studied by counting time periods within monumental observatories. An extraordinary map of astronomical time had emerged, still visible in the form and dimensions of the monuments. This earlier megalithic knowledge was very influential in forming the ancient world's myths of creation and of the gods.*

It therefore appears that megalithic achievements in astronomy and

---

*This is a subject of my earlier work, especially *Sacred Number and the Origins of Civilization*.

geodetics came via the same approach of manipulating numbers within exact measured lengths. Because of this approach, calculations could be performed without any notational arithmetic. Instead geometries such as the right-angled triangle and circle were used. Notational arithmetic uses symbolic marks detached from actual objects and identifies its numbers as concepts. In megalithic times there was no sense of number and its representation as we have now. "Numbers" were *counts,* stored using units of measure and transformed using geometry.

While feet came to dominate the geodetic metrology of Britain and Egypt, it was the inch, one-twelfth of a foot, that unlocked for me the power of the monuments found at Carnac in Brittany. The inch involved was apparently already standardized and is the measure still used today. It is close to the size of most thumbs when pressed onto a flat surface. Units similar to the inch are convenient when counting the days elapsed between astronomical events. Eventually such counting becomes a continuous calendar count as was the habit of the Maya, and within it all manner of astronomical events became somewhat predictable. Such astronomical regularity would be turned into gears by the classical Greeks, becoming simulators like the device attributed to Archimedes and called the Antekythera Mechanism, dated to roughly 200 BCE. Astronomically captured numbers were also stored within symbols, artifacts, and mythic story texts as a network of symbols mapping out astronomical and geodetic actualities. As an example the equator measured in inches gives a number found in post-Vedic myth: 4,320,000 inches, the number of "years" in the Hindu cosmology of the Yugas or "ages of the world" and the duration of one day of Brahma.[2]

If the equator were divided into 4,320,000 parts, then each would be 365.2422 inches long, the length of a solar year in day-inch counting, where each day is counted as one inch. Therefore the equator is exactly 4,320,000 solar years, the same length said to constitute only *a single day* of the creator god Brahma. This mythic Hindu reference clearly has a concrete relationship to early counting as developed by Carnac's astronomers.

Returning to the alternate division of the equator into 365¼ day-degrees, each 4,320,000 inches long, the method used to estimate this length would have involved a latitude such as that which runs through Brittany at latitude 48.19 degrees. At that latitude the circumference of

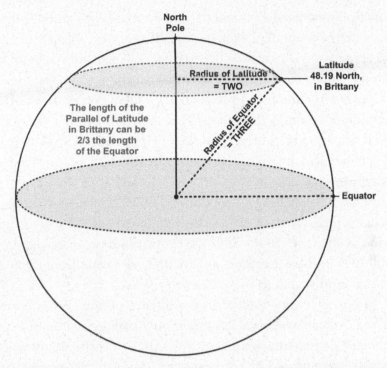

Figure 1.1. Diagram of cross-section of Earth showing how the ancients determined the Earth's radius (x) and circumference (2πx) specifically at latitude 48.19° north, in Brittany, where a simple right-angled triangle shows that the radius and hence circumference of that parallel and the equatorial radius must be related as a length 2 relates to a length 3.

the Earth is two-thirds of the equator, for the cosine of 48.19° is two-thirds, reducing a day-degree of that parallel of latitude from 4,320,000 to 2,880,000 inches long. In this one sees that the right-angled triangle is the key to trigonometry, and therefore geometry, soon gave the megalithic access to cosines and tangents as relative lengths when doing geodetic measurements.

RIGHT TIME, RIGHT PLACE, RIGHT CIRCUMSTANCES

Megalithic peoples appear to have chosen specific latitudes that enabled their metrological work to succeed. They did this in Brittany (5000–3000 BCE), where the latitude enabled them to study astronomy more easily (to be explored in chapter 2) and estimate the length of the equator to define a standard foot (as explained above). Megalithic peoples also chose a specific latitude between Stonehenge and Avebury in England (ca. 3000–2500 BCE), which enabled them to deduce the size of the Earth without a full geodetic survey. There is no evidence of similar preliminary works in the ancient Near East prior to the Pyramid Age, which employed megalithic metrology and its model of the Earth.

The ancient metrological system, in order to be the same everywhere, can only have evolved once in one region of the planet (just as the phonetic alphabet has evolved just once). Evidence for this development process is clear within the monuments of the early megalithic period in Brittany (5000–4000 BCE) and similarly in England (3500–2000 BCE), where it appears to have become a completed system ready to be communicated to the ancient Near East.

Metrology clearly evolved in the absence of arithmetic. It offered a method to obtain numerical results without needing the proto-mathematics developed by the Sumerians after 3000 BCE. When megalithic metrology was later adopted by the Egyptian, Sumerian, Babylonian, Semitic, and Greek cultures, it soon ran alongside the newer mathematical problem solving, which could manipulate numbers directly using symbols rather than lengths. Mesopotamian accountancy, counting the number of objects and adding and multiplying them, was a recording of *quantities* in symbols rather than lengths. Originally these quantity symbols were symbolic objects within a clay envelope then covered by both a summary of its contents and a seal. What we now call numbers are the summary without the seal, envelope, or miniature objects that were placed inside. Having two copies of the summary in the hands of each party proved a much more convenient contract for exchanges of goods and wealth.

Two thousand years before Sumer, however, the people of the megalithic in northwestern Europe were building a detailed understanding of astronomy and of the size of the Earth through measured structures that recorded long periods of astronomical time and modeled key radii and surface lengths in the Earth's dimensionality. The historical influence of the megalithic is only to be found in the measures within buildings, for the pyramid builders of the Near East and later of the New World employed the same megalithic units of length, summarizing and preserving megalithic knowledge in its native language of metrology. A history of the megalithic was never written, perhaps, because metrology was considered a sacred science, or possibly such a history did not survive the destruction of ancient libraries.

The ancient Near East is therefore not the beginning of the story leading to modernity. All post-megalithic buildings share an identical metrology and express ideas, even unknowingly, about the cosmos and its system of measures that originated in the megalithic period. Later pyramid-building cultures differed in latitude from the great megalithic sites and in their food production, treasuries, trading networks, and conquests that enabled the god-kings of Egypt, Assyria, and Babylonia to manifest historically. The new intellectual tools for building such empires included writing, mathematics, and metal working. It is these tools that displaced megalithic knowledge within a historical record focused on what those empires were achieving. We must remember that our history of the ancient world has only been completed in the last few hundred years through archaeological finds and decoded texts that added to the extant histories of ancient authors, revealing for example that the biblical Hittites were a major empire based in Turkey. So while history contains little record of the megalithic period, we can now recover historical information from its monuments because these embody the development of numerical invariants belonging to astronomical time and the size of the Earth.

Millennia of continuous patient adaptation prepared humanity for the leap in technical capability found in megalithic astronomy.

AN AWAKENING TO THE STRUCTURE OF TIME

How the Stone Age was able to awaken, to become the megalithic, can be seen in clues left by the late Stone Age. The Upper Paleolithic appears to have been counting using marks on bones for ten to twenty thousand years before the megalithic started.[3] Individuals scratched a tally of days on discarded bones, marking the elapsed time of the lunar cycle and, in at least one late example, of the solar cycle.

It is important to see that this Stone Age practice was not too different from the later megalithic counting of days. It was a step toward the megalithic discovery of a set of numerical coincidences within the time environment of the Earth. These coincidences would have been very thought-provoking to the people of the megalithic and actually run counter to our modern notion of a solar system created through natural accretion and therefore having no particular order. The discovered patterns, involving small numbers of days, months, and years, then formed the necessary bridge between the megalithic age and our more recognizable history of civilization, a bridge in which the world of time revealed itself as a numerically organized whole.

The Earth spins within a special kind of universe, where the outer form of things gives clues as to their function. This characteristic of the universe, often taken for granted, makes it possible to relate *at all* to how the world works. It is remarkable that the universe is *intelligible,* that is, many of its structures came to be known through discovering the use of simple numbers and geometrical forms. Our current understanding of the visible universe distinguishes us from the people of the Stone Age, when the emergence of numerical forms of intelligibility provided ways for the people of the megalithic to learn far more than we think, especially when using only simple technologies.

COUNTING WITHOUT NUMBERS

Today, counting is often done by verbalizing each successive number, but when a set of marks are made, what is "measured" is innumerate and mute unless it is given a number, as with the inches or centimeters on a graduated ruler. Our own numeracy makes any study of counting before numeracy difficult to appreciate. This problem of understanding the pre-numerate world without the filter of later developments, now intrinsic to our consciousness, was addressed well by Alexander Marshack and has led to the recognition that pre-numerate counting evidently had uses. Marshack found that studies of the moon's phases on bone tallies were remarkably widespread during the Upper Paleolithic era in France, centered around a northern latitude of 45 degrees (south of Carnac).[4]

Bones from different epochs of the Upper Paleolithic held carefully scribed marks that appear to be a pre-arithmetical counting of the lunar month. The marks resembled a count on a prison wall, however, Marshack went to considerable lengths to *avoid* assuming that such groups of marks indicated any comprehension of number, which many other interpreters had immediately sought to do. He found it more plausible that these productions were just counting the length of a synodic month, the "phases of the moon" formed by one or more months of day-marks, often grouped in different but equally practical ways. He notes that because these productions were for personal use, variations in technique could be expected and these often displayed a creative and inventive working of the bones.

For example, Marshack proposed that practical, pre-numerate observers used the disappearance of the moon (the new moon) and its subsequent reappearance as an evening crescent as a clear sub period of 2–3 day-marks, so that two invisibilities and one set of visible phases might sum to 32 or more day-marks in chunks of 28 and 4 or similar. There was no concept of "finding the number of days in a synodic month," as the megalithic did, but rather an open-ended exploration through the recording of conveniently observed periods between visually significant moments in the moon's appearance. Thus they created a tool for remembering what had been noted, using a *notation* instead of a numbered set of marks. As Marshack continued in this vein, evidence of longer counts and a virtual certainty that lunar months were being notated led him to find evidence for a later tradition of longer, year-long counting. These may have been due to an increased interest in the seasons as relevant to early agricultural practices.

After publishing *The Roots of Civilisation* in 1972, Marshack was informed that a complex notated object dated to roughly 10,000 BCE had been found at the Grotte du Tai in France: the Tai Plaque. It would be twenty years before its meaning became properly evident to him:

> The Tai composition, dating to the terminal Magdalenian, is in many respects the most complex single artifact to come from the European Upper Palaeolithic. The present analysis has for that reason been the most thorough yet essayed for any composition from that period.[5]

The first and perhaps only surviving device to record a lunar calendar period of twelve months, the Tai Plaque spans about three years of non-arithmetic counting, shown as six half-years and recorded in the usual

Figure 1.2. The Tai Plaque. The primary face of the engraved rib fragment from the Grotte du Tai. Note the two bars connecting the verticals on the right of the main face. From Alexander Marshack, "The Tai Plaque and Calendrical Notation in the Upper Palaeolithic," used with permission.

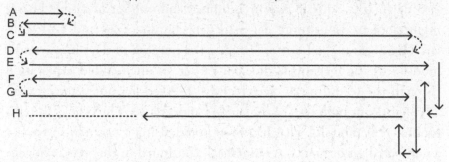

Figure 1.3. The Time Factoring. Schematic model of the "boustrophedon" or serpentine mode of engraving the sequence of lines. Adapted from Marshack, "The Tai Plaque and Calendrical Notation in the Upper Palaeolithic," figure 2.

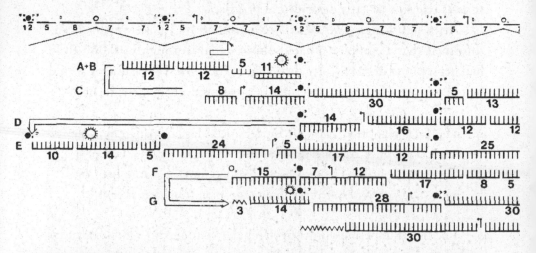

Figure 1.4. The Tai calendar. The sets in the Grotte du Tai composition laid out sequentially against an accurate lunar model. Possible solstitial positions are indicated.

"serpentine" or *boustrophedon* sequence. There is evidence of a significant deviation within the second year of notation where the count appears to have moved at a right angle in order to await a summer or winter solstice event. This deviation naturally extended the second year count to produce a *solar* year count rather than a *lunar* year count.

One can see that if counting in days was to lead to numeracy, the grouping of small numbers of marks and the naming of these groupings would draw attention to numbers as useful and spur the rise of numeracy in human consciousness. The counting of months in the solar and lunar years carries the numbers 12 and 13. These small numbers operate more like shapes in that sometimes they fit with each other and sometimes they do not. This complements the role of numbers in geometrical patterns, widely practiced as decoration in the Upper Paleolithic. Geometrical decoration contains numerical relationships within a given boundary and an orderly and balanced repetition, expressive of pure number.

According to Marshack's meticulous scrutiny using a microscope, the first four lines on the Plaque contain 356 marks. This appears to show the length of a lunar year of 12 full synodic months as being about 354 days long. The next two lines sum to 370 marks, but we know this includes some belonging to other lunar years. Tallying between the key days in the year, when the sun appears above the same position on the

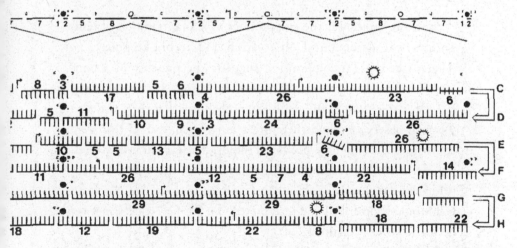

From Alexander Marshack, "The Tai Plaque and Calendrical Notation in the Upper Palaeolithic."

horizon, corresponds to 365 day-marks, a solar year. The third year is incomplete due to damage, lack of space, or the completion of aim in making the Plaque. It is not the number of marks that signifies what was to be learned but the perceived length of sets of marks, relative to each other, that marked duration.

The lengths of the Tai Plaque's lines of marks clearly belong to lunar and solar year lengths in days, observing the moon's general pattern of phases and its disappearance when close to the sun. Because the day-marks are about half a millimeter apart on a bone only 88 millimeters long, this plaque can be compared to Carnac's Quadrilateral where the duration of one day was marked by exactly one inch. As a megalithic monument the Quadrilateral similarly recorded two lunar years of 12 months and one lunar year of 13 months, making 37 months in all, which nearly equals 3 solar years. While the Quadrilateral is a count at fifty times the scale of the Tai Plaque (which had to be portable art), the Plaque is very close to being the natural precursor to the Quadrilateral (see figure 1.5, showing the southern kerb).

Marshack saw the Plaque as representing the high point within an independent European tradition of astronomical observation, a tradition that had evolved during the Upper Paleolithic and that seems directly relevant to the megalithic astronomy being practiced at Carnac, its natural precursor. He said:

> . . . the Tai notation is the end-product of a long tradition of non-arithmetical astronomical observation and record-keeping. . . . Analysis of the Tai composition suggests the presence of a non-arithmetical observational lunar "year" in Europe at the end of the last Ice Age, c. 10,000 BC, with the probability that there was also solstitial observation and therefore a lunar/solar year. If this were confirmed, it would have profound importance for our understanding of the 25,000-year cultural development of the Upper Palaeolithic. The regional "explosion" of culture and symbol has been voluminously described and interpreted, but has nevertheless not yet been adequately explained. . . .
>
> The presence of such a year structure in the terminal Palaeolithic would argue for an indigenous European observational astronomy in

the Neolithic and Bronze Age that was not derived, either in structure or mode, from the astronomies of the classic Mediterranean agricultural societies. *The origin of this West European megalithic astronomy has been one of the principal problems confronting the new discipline of archaeoastronomy.* [my italics]

The alignment astronomies of the West European Neolithic suggest that an observational lunar/solar calendar such as that represented in the Tai notation may have provided the base for the observational lunar/solar "calendars" of the megalithic cultures. This European observational astronomy seems not to have been derived from the Near East or Egypt, but from the Palaeolithic tradition documented by the Tai notation.[6]

MEGALITHIC NOTATION USING DAY-INCH COUNTING

Marshack invites us to search for the bridge between Upper Paleolithic and megalithic and this can be seen at Carnac, where the size of each day-mark was standardized to a larger and more reliable unit of length, the inch. A fixed length, the *inch*, made counts repeatable and comparable. This approach inevitably made for larger, non-portable counts, no longer restricted to the counting found on "portable art" objects. The lack of signs of settlement at Carnac suggests that the megalithic sites were set apart for their special purpose. Increasing levels of regional settlement could have supported sites devoted only to the conduct of astronomical counting and observation.

At Carnac's Quadrilateral the stones clearly represent lunar months. Stones 36 and 37 at the end (see figure 1.5) of the count from stone 1 (left) correspond to the months in three lunar years and three solar years, respectively. This *symbolic* counting of 36 and 37 lunar months, as the 36th and 37th stones, was underscored by an *exact* day-inch count equaling the number of days in three lunar years (1,063 days) and three solar years (1,095.75 days).

A standardized unit of length gave new potential to the aggregated time periods of Stone-Age tallies. Lengths could be compared, such as when two ropes hold the counting length of two time periods

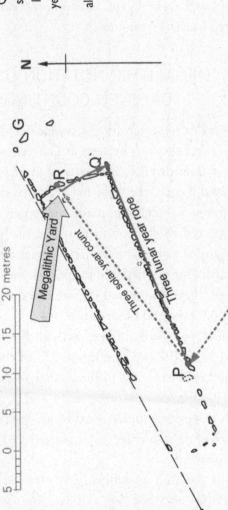

Figure 1.5. The Southern Kerb of Le Manio Quadrilateral was counted in day-inches from the start of stone number 1. Three lunar years of 36 lunar months ends at stone 36 while three solar years of 37 lunar months ends with stone 37. Thus a nominal month count using kerb stones runs alongside an exact one-inch-to-one-day count over the period of three solar years.

Figure 1.6. Thom's site plan for Le Manio Quadrilateral with four key points: The line P to R represented a count of three solar years, starting at P and ending at dressed edge of stone R. This line could be extended to the groove in stone G, symbolically marking the midsummer solstice sunrise.

in day-inches. These, laid out from a common pivot stake P (see figure 1.6) so as to form a right-angled triangle that would make the Quadrilateral's southern kerb a baseline of the shorter length (36 lunar months *to stone 36* and point Q') and the longer time length (37 lunar months *to stone 37* and point Q") ending the southern kerb's major stones where a three solar year count would end.

To have lengths that numerically represent time within ancient monuments is not altogether new, but these have never been seen as a primary means for astronomical measurement. Instead they have been viewed only as a symbolic presentation of time, an "after the fact" codification based upon the knowledge of time periods or recording a seasonal or religious calendar. In day-inch counting the symbol and the measurement are unified as a directly useful representation, still acknowledged unconsciously when time is talked of as having a length, as in "How long did it take?"

It may be difficult to accept the premise that our modern inch measure was actually employed seven thousand years ago as an improvement on the notation of Paleolithic bone tallies, but this unexpected view of the inner structure of monuments reveals the purpose of these monuments in a new and consistent way. It shows that day counting was performed, with measurements clearly in day-inches, an inch that is one-twelfth of an English foot.

Metrology is therefore a vital forensic tool for studying these megalithic monuments. It has undoubtedly been ignored because of the questions it raises, such as the seemingly anachronistic transmission of historical measures within them. However, key sites like Carnac in Brittany (chapter 3) and Stonehenge and Avebury in Britain (chapter 6) are a metrological codex, analogous to the Rosetta Stone, which enabled Egyptian hieroglyphs to be read for the first time. We are in the process of deciphering ancient languages that were mathematical rather than verbal. This requires a kind of thinking that we are not used to. We should remember that for the most part, advances in any field are accompanied by amnesia of what came before.

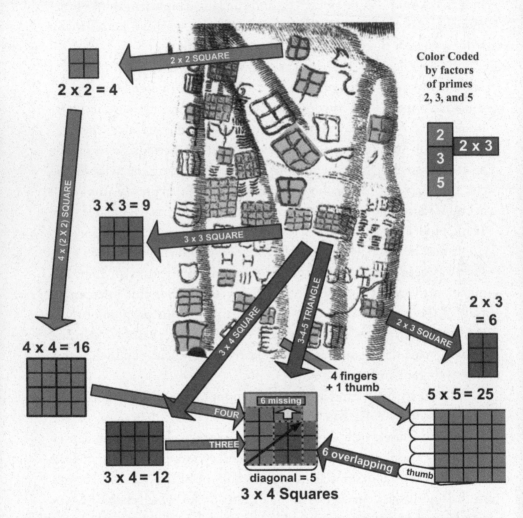

Figure 2.1. Etching by Debric of paintings in the rock shelter of Cachao da Rapa (Portugal), published in 1734 in Contador de Argote's *Memorias para a Historia Eclesiastica do Arcebispado de Braga*. These rectangles and parallel lines were painted in red, burgundy, and a rare very dark blue. It is virtually outdoors because they are not in a real shelter. Post-Paleolithic and probably early Chalcolithic (i.e., 3500 to 3000 BCE) and associated with the spread of the megalithic phenomenon to most of the Atlantic region in Europe.

2

THE TRANSMISSION
OF THE SQUARES

THE TWO GREATEST empires of the ancient Near East each built notable monuments and developed geometrical and arithmetical systems. In the Nile valley of Africa, the Egyptians are best known for their large pyramids built a few centuries after their meteoric cultural emergence under the earliest pharaohs. Their pyramids and temples employed sophisticated techniques based upon metrology. The Sumerians emerged a couple of centuries earlier, in the plain of the twin rivers Tigris and Euphrates, at the other end of the Fertile Crescent (see figure 2.2).

These two descending horns of the Fertile Crescent gave rise to an Egyptian sacred canon and Sumerian secular innovations. The Sumerians evolved the city-state as an independent trading unit, where a new arithmetic and cuneiform documents managed society's utilitarian needs. Cities were seen as the manifestation of a city god, each king being the city god's representative. Sumerian innovations were defining the economic communities of the future while the Egyptians were acting out a sacred canon already perfectly formed in Eternity and with little need for change.

Archaeological evidence of the precursors to the Sumerians are identifiable one to two thousand years before their civilization emerged in 3000 BCE. There is not much evidence in Egypt that predates the

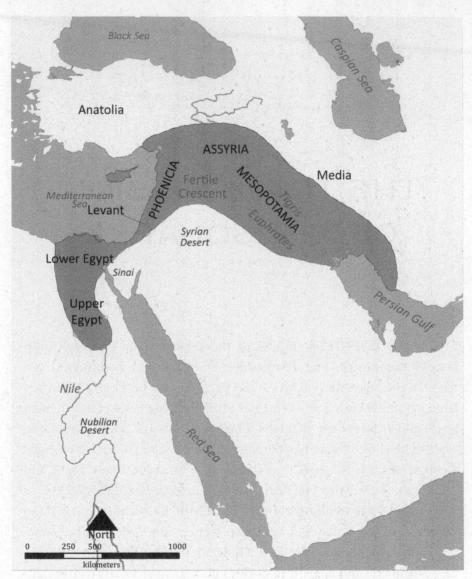

Figure 2.2. The Fertile Crescent at its greatest extent, made up of three great rivers, Nile, Tigris, and Euphrates, plus the eastern coast of the Mediterranean Sea.

first Egyptian dynasties' building of the greatest stone building of the ancient world the Great Pyramid. However, if one turns to northwestern Europe, one can see how sacred science, geometry, and metrology had been developing around the astronomical megaliths near Carnac for at least two thousand years before the pyramids were built. By the time the Great Pyramid was being built in Egypt (2560 BCE), another quite different masterpiece was being completed in southern England, the Sarsen Circle at the center of Stonehenge. While not very large, the Sarsen Circle holds information very similar to that memorialized within the design of the Great Pyramid, geodetic information about the size, shape, and surface distances of the Earth. It is clear the builders of Stonehenge were using the same metrological system employed within the Great Pyramid. This metrology evolved over more than a thousand years in Brittany, appearing in Ireland, in southwestern Britain, and at the beginning of the Pyramid Age in Egypt. So why did it need to be transmitted to Egypt?

By 3500 BCE the megalithic peoples of northwestern Europe had come to the end of their active learning about astronomical time and had become interested equally in the Earth itself. Their study of the Earth probably went through three stages of deepening understanding:

1. The planet could only be a sphere that was spinning. It had a north pole that could be seen in the northern sky, surrounded by stars circling eternally in a counterclockwise direction. Initially, the Earth would be assumed to be a perfect sphere. To estimate the length around the Earth, the length of the observer's parallel of latitude could be quantified as being a known fractional ratio of the equator itself.

2. Because the spin of the Earth causes the equator to grow, when traveling north over two or more degrees of latitude, the length on the ground attributable to a single degree of latitude gets larger. When discovered, this revealed that the polar radius of the Earth was shorter than the equatorial radii; but by how much?

3. To know the polar radius requires that the shape of the Earth be known. Like the orbits of the planets, its shape is called an ellipse. The known major axis of the Earth's ellipse is the equatorial diameter and its unknown minor axis is the polar diameter. The length of the polar axis relative to the equatorial can normally only be discovered by conducting a survey in two different regions of latitude.

This period after 3500 BCE must have been when the megalithic was measuring degrees of latitude in England because, due to the degrees of latitude encoded in the Great Pyramid of Giza (completed by 2560 BCE), geodetic surveys in two different regions of latitude would have to have been completed before the Great Pyramid could even be designed.

If a survey of the Earth occurred in England and Egypt in 3000 BCE, this explains how the two most famous geodetic monuments came to be built by 2500 BCE. Stonehenge and the Great Pyramid both used the same metrology to express different versions of the same model of the size and shape of the Earth. Egypt was an ideal "other place" for the megalithic peoples to conduct their geodetic survey: the Nile runs from south to north over many degrees of latitude, and its valley is cut into the rock of the western Sahara, providing raised observing platforms on either side.

Other types of exchange would have taken place during such a period of geodetic cooperation. Megalithic knowledge and techniques would have been shared if only to achieve the survey's objectives, and these could have, at the very least, helped to define the Egyptian sacred canon. Egypt had a bafflingly stable pattern of pharaohs, rituals, art, and science, and this would come to include a geometry of multiple squares called *the Canevas* by Schwaller de Lubicz.[1] As we will see, multiple squares were uniquely relevant in establishing an exact science of astronomy for the megalithic period at a specific latitude in France.

The earliest written source we have on the properties of multiple squares was recorded around 1500 BCE in the Rhind Papyrus, a rare surviving mathematical text from Egypt's New Kingdom. It documents parts of a geometrical tradition, relevant to the building techniques found from 5000 to 4000 BCE near the present-day town of Carnac (Brittany, France). Such an interest in squares was also arguably latent in the decoration of objects from the Upper Paleolithic. So what caused the megalithic love affair with multiple squares at Carnac in Brittany that may have been communicated to Karnak in Egypt during a later period of geodetic cooperation between Egyptians and the visitors from Hyperborea (as northwest Europe was known)?

FINDING THE PERFECT PLACE

The astronomers of Carnac aligned their monuments to observe the extreme horizon events of sun and moon, when rising and setting at that given latitude, the days between extremes being counted. The people of the megalithic must have inherited a material cultural background in which there was time to question the structure of the world and a willingness of regional communities to help what was a cultural quest and not a practical necessity. It took more than two thousand years to complete their quest for knowledge of the world, longer than the lifetime of any cultural mode known to history, apart perhaps from

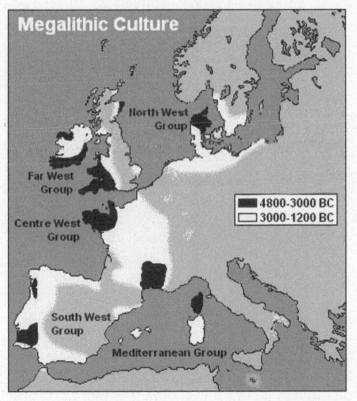

Figure 2.3. Map of the Megalithic Regions of Western Europe. Carnac is on the south coast of the Center West group (4800–3000 BCE) in Brittany. The Far West group was strongly related in its practices and now appears to have been the same cultural group of megalith builders that developed astronomy, metrology, and knowledge of the size of the Earth.

the Stone Age, which launched the megalithic metrological quest.

This culture's discovery, in its early stages, was that small whole numbers organized astronomical time. Also important was how the latitude at Carnac made astronomy easier, a situation not reproduced at any other latitude. Furthermore, this latitude was only a few degrees north of the nexus of earlier Stone Age sites in southern France, including the famous painted caves from the Upper Paleolithic. Someone must have realized that the latitude 47.5 degrees north would hold special promise for the conduct of megalithic astronomy, perhaps after living at that latitude for a few years.

In southern Brittany the extreme solar and lunar risings and settings (at solstice or at lunar standstills respectively) had unique *geometrical* properties. The diagonals of very simple rectangles could predict the limiting angle within which lunar and solar events could happen on the

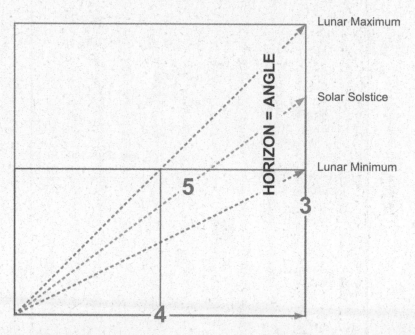

Figure 2.4. The simplicities of alignment found at Carnac regarding the extremes of the sun and moon follow, for the moon at maximum standstill, the diagonals of a single-square (45 degrees) and at minor standstill a double square (26.565 degrees) while, for the sun at either solstice, the 3-4-5 triangle's smaller angle of 36.8 degrees, all relative to east (when rising) and west (when setting). This shows a single quadrant of observing, the observer at bottom left, east to the right, and north upward. Figure 2.5 shows all four quadrants.

horizon, from their rising relative to east and setting relative to west. This provided a coherent starting point for horizon astronomy at any observing site. For this reason these geometries are ever present in the design of Carnac's megalithic structures and in their alignment to other structures, generating a landscape of megalithic interalignment related to the extreme positions of the sun and moon upon the eastern and western horizon and the boundaries for the behavior of the sun and the moon.

Looking at figure 2.4, the extreme solstitial sunrise at Carnac created the simplest whole number triangle, the 3-4-5 triangle (with the FOUR side lying due east and the hypotenuse of FIVE pointing to the rising sun) twice a year. As shown in figure 2.5, in summer the THREE side ran north and in winter south. In between these points lies the Equinoctal point, marking the spring and autumn equinoxes when the sunrise is exactly east.

There are two points, called lunar nodes, where the moon's path crosses that of the sun. These nodes establish the extreme range of moonrises that occur within each lunar orbit of $27\frac{1}{3}$ days, a range that waxes and wanes over 18.6 solar years. The sun's extremes (the solstices) mark midwinter and midsummer within each solar year. The nodal variation of the moon's extreme range grows and it can exceed the angle of the sun's solstitial extremes over the year. Eventually the moon's range reaches a maximum standstill, whereupon, for some months, its orbital extremes on the horizon, to north and to south, were aligned with the diagonal of a single-square (to east for moonrise and to west for moon set). Such maximum standstills occur every 18.6 years (or 6,800 days). This lunar nodal cycle is the result of backward (or retrograde) motion of the moon's orbital nodes around the entire ecliptic. Thus the two diametrically opposite crossing places of the moon's orbit, in front of the sun's ecliptic path, systematically create sequences of eclipses over about eighteen to nineteen years. The ascending node is where the moon crosses to being north of the sun's path while the descending node is where it crosses to being south of the sun's path.

The sun's path is also itself angled due to the tilt of the Earth so that the sun crosses the celestial equator at the two equinox points, points that move around the ecliptic in another retrograde cycle called the Precession of the Equinoxes, which takes 25,920 years to complete.

Once seen clearly it would have been natural to relate the Precession of the Equinoxes with the moon's nodal cycle because both are movements due to a non-alignment of (a) the moon's orbit to the sun's path and of (b) the axis of the solar system relative to the polar axis of the Earth.

The lunar nodes were very important to megalithic science because eclipses of the moon in particular reveal a lot about the structure of time on Earth. Megalithic astronomy would come to associate the paths of celestial objects and their cycles with one's journey after death, which was thought (in later cultures) to take place along these non-aligned paths of sun and moon to reach a galaxy that in 5000 BCE descended vertically to touch the Earth at the points of spring and autumn equinox.[2] This may be why monuments like chambered tombs are found to point to the sun or moon in their extreme north or south alignments. In

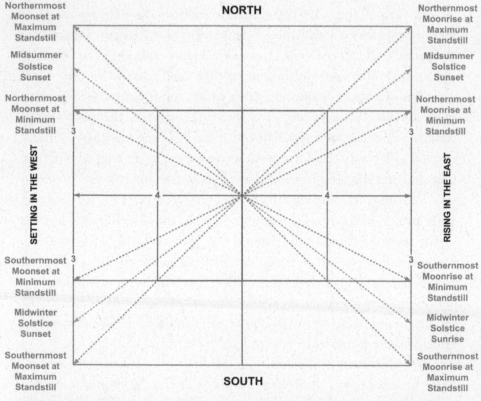

Figure 2.5. The simple geometry of extreme sunrises for each quadrant of an observation point (the center of the square) in Carnac during the fifth millennium BCE.

contrast, the Precession of the Equinoxes came to be called—by Plato's time in 500 BCE—the Great Year because it resembles the pattern of the solar year but over an epic period of 25,920 solar years.

Between the maximum lunar standstills every 18.6 years, the moon's orbit becomes less extreme in its range on the horizon. When no longer exceeding the sun's extreme position at solstice sunrise, the moon approaches its minimum standstill, where it aligned with the diagonal of a double-square (a two-by-one rectangle).

To summarize: the moon's extreme range alternates between a minimum and a maximum angle in the east every 9.3 years (3,400 days). These angles of minimum and maximum standstills approximated well to the diagonals of a double-square and of a single-square, respectively. During each orbit the moon makes of the Earth, the extreme angular range of moonrise is either decreasing or increasing between these two standstill positions on the horizon.

Carnac became a center for megalithic astronomical studies because of its latitude, which enabled these specific geometric alignments for the sun and moon on the horizon. These alignments made Carnac's horizon astronomy extremely intelligible since a simple ground plan, like that shown in figure 2.5, could be laid out to indicate where the angular limits of sun and moon could be expected on the horizon over the year and the moon's nodal cycle. Moving to the latitude of Carnac gave astronomy a simple geometrical form. The selection of this latitude was surely empirical, made most likely by a Neolithic culture southeast of Brittany.

A BIGGER COINCIDENCE CONCERNING SQUARES

Even more surprising than these horizon geometries is their strange compatibility with the structure of the solar and lunar cycles. When time was counted to form two lengths equal to the solar and lunar years, these two lengths were found to form the longest side and diagonal of a four-square geometry (a four-by-one rectangle). This geometry was found to be more accurate than simply counting days and represents the very first device for predicting the location of the sun and moon on the ecliptic. The solar year and the eclipse year (the time taken for the

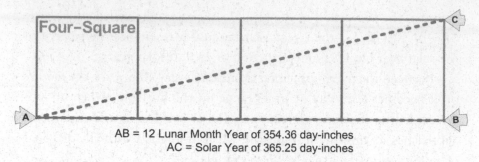

AB = 12 Lunar Month Year of 354.36 day-inches
AC = Solar Year of 365.25 day-inches

Figure 2.6. The solar and lunar years, as a ratio, accurately follow the base length relative to the diagonal length of a four-square rectangle.

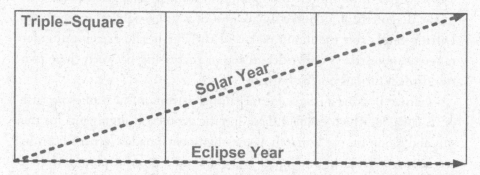

Figure 2.7. A triple-square approximates the ratio, in relative lengths, of the solar and eclipse years within its base and diagonal.

sun to again sit on the same lunar node) can also be represented by the longest side relative to the diagonal length of a triple-square.

Therefore, in the space and time of Carnac, we find the multiple squares for 1, 2, 3, and 4 were all present. Astronomy followed this geometrical theme of multiple-squares at the latitude of Carnac, using the numerical sequence of one-, two-, three-, and four-square rectangles. The geometrical relationships of eclipse year, lunar year, and solar year could well have been discovered wherever day-inch counting first originated, their ratios seen as conforming to the three- and four-square rectangular geometries. Upon arriving at Carnac, the complementary geometries of solar and lunar alignment there must have given great significance to the special nature of these multiple square geometries and to the whole number ratios arising within them. In modern times this

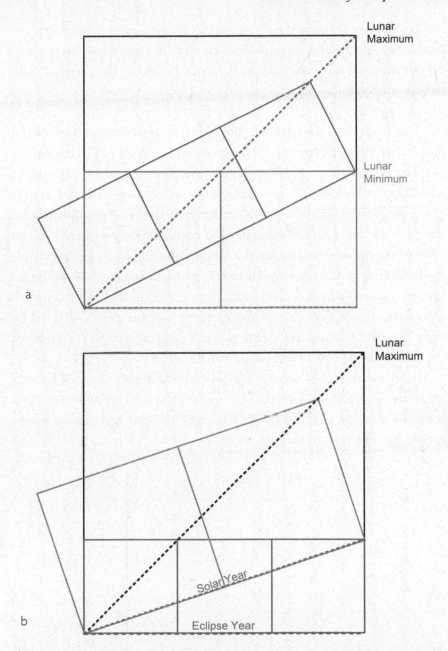

Figure 2.8a and 2.8b. (a) A single-square's diagonal angle can be achieved by building a triple-square on the diagonal of a double-square (top), or (b) building a double-square on the diagonal of a triple-square (bottom). The reason is that the difference between the single-square and double-square diagonals is the diagonal angle of a triple-square, so the latter can perfectly "bridge" the angular difference between the single- and the double-square.

correspondence between the geometries of astronomical periods and the geometries of sun and moon would be seen as an extreme astronomical coincidence, a miraculous coincidence in which the number field strangely linked the celestial periods and horizon observations of the sun and moon.

This leads us to the question: Do multiple squares have some special properties? The first four multiple square rectangles do indeed have the property that their diagonal angles are complementary, in that placing a further multiple-square geometry onto an existing diagonal arrives at the diagonal angle of another multiple-square. For example, figure 2.8 on p. 33 shows how the double- and triple-square geometries add to 45 degrees, the diagonal angle of a single square. This is clearly visible in the design of Carnac's monuments, which often stacked multiple-squares in combination within monuments, to indicate both the invariant ratios of time periods and the extreme horizon events affected by time. Figure 2.9 shows two triple-squares "adding up" to the angle of the solstice angle of a Pythagorean 3-4-5 triangle.

In this marriage of compatible geometrical forms, a type of astro-geometric grammar became possible within Carnac's monuments when counting time or aligning to the limits of the sun and moon. These simple geometrical archetypes were compatible with each other and

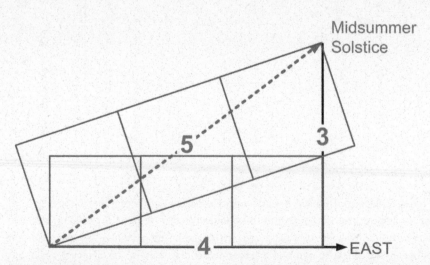

Figure 2.9. Pythagorean 3-4-5 triangle can be achieved by similarly stacking a triple-square on another triple-square's diagonal.

became symbolic of those celestial luminaries. Multiple-squares offered a perfect medium for astronomical work prior to the exact sciences of Babylonia in the third millennium.

The Role of the Seven-Square

The diagonals of the double-square and triple-square add up to the diagonal angle of a single-square (figure 2.10), while the diagonals of

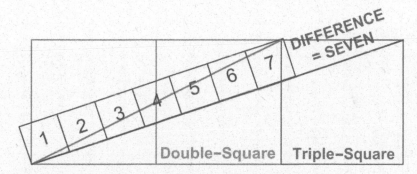

Figure 2.10. The differential angle between a double-square and a triple-square is the diagonal angle of a seven-square.

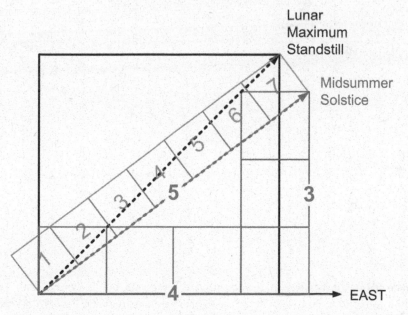

Figure 2.11. The difference between the 3-4-5 triangle and a single-square's diagonal angle is the diagonal of a seven-square. This arises because, as in figure 2.8 on p. 33, the seven-square is the differential angle between a double-square and a triple-square.

two triple-squares add up to the 3-4-5 triangle's smaller angle (figure 2.11 on p. 35). This points to there being something systematic about the difference between a triple-square's and a double-square's diagonal angles and this is found to be the diagonal angle of a seven-square, which forms a bridge between a single-square's diagonal and *either* of the 3-4-5 triangle angles, as seen in figure 2.11 and in the later section on Egyptian *Canevas*.

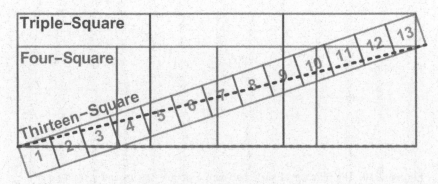

Figure 2.12. The difference between the diagonals of a four-square and a triple-square is the thirteen-square's diagonal angle.

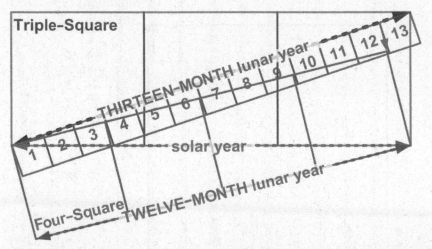

Figure 2.13. The baseline of the four-square represents the 12 lunar month year relative to the solar year, while the triple-square represents the ratio between the solar and eclipse years as well as the 13 lunar month year relative to the solar year. This has a direct astronomical relevance to the use of three-square and four-square structures where a triple-square's diagonal can refer to a lunar year of 13 months.

The Role of the Thirteen-Square

If the attention then turned to the third differential angle, between the diagonals of the triple-square and the four-square, this would easily be discovered as being the diagonal angle of a thirteen-square, as is demonstrated in figure 2.12.

The diagonal of the triple-square is not only the invariant angular ratio between the solar and eclipse years, but it expresses the same invariant ratio as between a 13 lunar month year and the solar year. Since the four-square baseline represents the more familiar 12 lunar month year relative to the solar year, the triple-square and four-square can be rearranged as in figure 2.13 so as to relate the solar year held between the two types of lunar year.

Carnac's megalithic monuments employed most of the above angular manipulations using multiple squares, which were therefore known to astronomers at the time of the monuments' construction. The fact that the two multiple squares relating astronomical time, the three- and four-squares, should be separated by the differential angle of a thirteen-square geometry is astronomically significant. The 13-month year is a longer lunar year that relates to both the

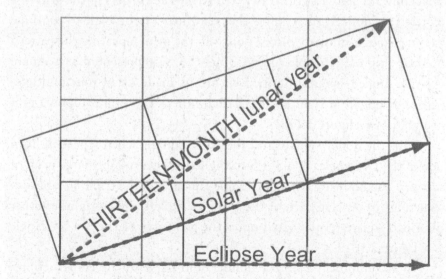

Figure 2.14. Triple-squares add up their inner angles to equal exactly the angle between the FOUR and FIVE sides of a 3-4-5 triangle or, seen alternatively, the angle of a four-by-three rectangle's diagonal.

solar year of 365.2422 days and the eclipse year of 346.62 days. This placed the counting of time within a natural geometrical framework.

Figure 2.14 on p. 37 shows how counting can populate this geometric framework. Thirteen months relates to the solar year as the diagonal of a triple-square relates to its triple side length. In just the same way, the solar year relates to the eclipse year as the diagonal of a triple square relative to its triple side length. That is, *two triple-squares bridge the ratio between the 13-month lunar year and the eclipse year.* When drawn on an east-west baseline, the solstitial sun will shine down the 13-month diagonal.

THE EGYPTIAN CANEVAS (2500–1500 BCE)

The multiple squares found within Carnac's monuments were made to align to horizon events or to record times between celestial events as lengths. This evolved a building technique based upon a grid of squares, or *canevas,* where a diagonal was not amenable to any whole number representation, appearing, most likely, as an irrational square root. However, there is a kind of rationality found in these diagonals as they travel onward across a grid. They will strike the corners of the grid in a metered fashion. At such points they can turn a right angle to continue passing though similarly placed points in the grid, repeating the pattern of their original multiple square. This can be represented as a square with an angled square inscribed within it, made up of four multiple-square diagonals at right angles to each other's end points, upon a grid of squares (see figure 2.15).[3]

Both the double-square and the triple-square, when repeated, form one of the two angles of a 3-4-5 triangle. Using a suitably large grid, or *canevas,* and allowing both the double-square and the triple-square constructions to develop each of these angles (to left and to right) another inscribed square results based upon the geometry of the 3-4-5 triangle (see figure 2.16).

To accommodate *both* the double- and triple-square constructions, this grid requires a doubling of the grid units, allowing 8 grid points on the left, to contain the double-squares, and 9 grid points on the right, to contain the triple-squares. The resulting "tumbling square" of figure

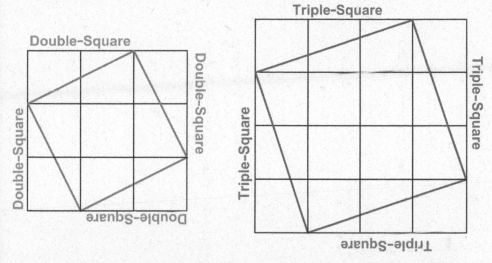

Figure 2.15. The Canevas in which the irrationality of diagonals can be bypassed for the purpose of building, using only whole number lengths. After R. A. Schwaller de Lubicz, *The Temple of Man*, p. 236, fig. 86.

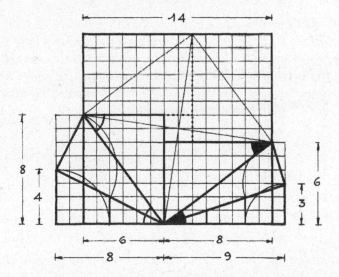

Figure 2.16. The Canevas of the 3-4-5 triangle, formed using two double-squares (left) and two triple-squares (right). Illustration from *The Temple of Man*, p. 236, fig. 87.

2.16 requires only 14 of these grid squares to contain its height and width. (The tumbling square has side length exactly 10 squares and an area of 100 because the 3-4-5 sides are doubled up.)

As already mentioned this production belongs to the pharaonic

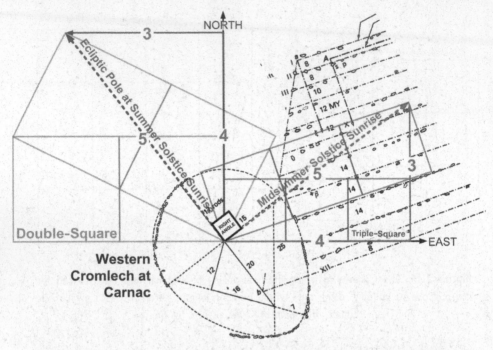

Figure 2.17. At Carnac the ecliptic pole was located in 5000–4000 BCE using two double-squares and, at right angles to it, the solstitial sunrise found using two triple-squares, all related to the greater and lesser angle of a 3-4-5 triangle, respectively, relative to west for the ecliptic pole and east for the solstice. The shown construction for summer solstice sunrise follows the major and minor axes of the stone circle at Le Menec called the western cromlech.

Egyptian mathematics of the *Rhind Papyrus,* a rare physical survival from around 1500 BCE. Just this one demonstration, of an Egyptian geometrical solution using multiple squares, exactly describes the situation as it was for the astronomers at Carnac in Brittany. This geometry represents both the solstice angle and that of the ecliptic pole in the circumpolar skies of Carnac. These angles are always perpendicular and offset from the cardinal directions east and north, respectively, by the smaller angle of the 3-4-5 triangle. The location of the ecliptic pole at solstice is therefore at *the other* and larger angle of a 3-4-5 triangle relative to west. The similarity to the Egyptian "example" is strikingly seen at Le Menec (figure 2.17).

Schwaller de Lubicz summarizes the ancient Egyptian approach to geometry as follows:

The mentality of the Ancients is geometric (functional) and, in Egypt, it always refuses the scholarly form that substitutes the mental concept for the graphic means. It remains faithful to the fractional system, refusing a decimal system that necessarily moves away from geometry. The link between fractional calculation and geometry is made by the trigonometric notation of 1:n. This synthesizing notation allows us to place canon, architecture, and calculation on a sort of "backdrop" that we call canevas, the grid pattern used by the temple builders. We might be tempted to see ordinates and abscissas in the canevas, but this would not be "pharaonic."[4]

This is exactly the kind of approach one would expect from a tradition that began before the development of the later "scholarly" mathematics in Babylonia. In the canevas the Egyptians appear to have had a megalithic approach to geometrical transformations even in the second millennium BCE. From this, one can perhaps glimpse into the intellectual traditions of the megalithic in Brittany. The Egyptian use of a pragmatic megalithic approach to geometrical calculation was not purely for religious art but defined their "canon, architecture, and calculation."

It seems plausible that the Egyptians inherited their metrology and geometrical method of calculation based upon geometry from the megalithic people of northwest Europe. Only a period of collaboration could have led to the Great Pyramid and the Sarsen Circle within Stonehenge both contemporaneously being models of the Earth's size and shape.

Figure 3.1. Alignments at Le Manio's Quadrilateral mark the end of various types of year with the Menhir Geant in the background. Photo by author during survey.

3

MEGALITHIC REVELATIONS AT CARNAC

THE "COLLISION" OF the multiple-square geometries of time periods and of horizon events at Carnac due to its latitude confer an understanding of celestial events far simpler than our own. Through its single-square, double-square, and four-by-three-square rectangles (of the lunar and solar extremes on the horizon), the latitude enabled the lunar and solar day-inch counts to fit within the four-square geometrical framework. This would have avoided the intellectual prerequisite of inventing the right-angled triangle as such. Instead, the development of the right-angled triangle would have naturally followed on from the simpler utility of multiple-squares.

Right-angled triangles are important because they enable any two time periods to be compared as lengths, to visually display any invariant ratio whose signature is the angle formed within that triangle. Since two identical right-angled triangles make up each multiple-square rectangle, forming such rectangles marked a very auspicious beginning for the subsequent discovery of the trigonometrical functions within the side lengths of right-angled triangles. Trigonometry was critically important for land surveying in the later megalithic and, after this, the development of our early mathematics and astronomy in the ancient Near East.

THE MULTIPLE-SQUARE GEOMETRIES OF LE MANIO

The astronomy at Carnac counted astronomical periodicities and therefore sought anniversaries in which two periods could find closure, from an observable starting event to a near-identical ending event. At Le Manio, a stone kerb called the Quadrilateral appears to have been designed after noticing that three lunar years, 36 months, was approximately one lunar month short of three solar years, which include just over 37 lunar months. This means three lunar years forms approximately an N:N+1 ratio to the solar year, where N=36, making the solar year *approximately* 37/36 times as long as the lunar year.

Looking at the two year-lengths as two ropes counted in days using inches: the difference between the three-solar-year count in day-inches, 1095¾, and the three-lunar-year count, 1063⅛, is 32⅝ day-inches, the

Figure 3.2. Le Manio from the northeast of the Quadrilateral (right) and looking down to the Menhir Geant to the south (left).

length of the average lunar month (of 29.53 day-inches) plus a little more than three inches. This length is the megalithic yard, found in the twentieth century by Alexander Thom when he surveyed Britain's megalithic monuments. Thom, a leading aeronautical scientist of his day, spent most of his spare time between 1937 and 1961 surveying British megalithic sites.[1] Thom helped define what was a neglected area of study and found evidence of a "megalithic science" in Britain. To test his hypothesis, he was invited by the editor of the scientific journal *Nature* to test whether Carnac's famous monuments had similar characteristics and hence, part of the same megalithic culture Thom had identified in Britain. He found many similarities to British monuments, though Carnac also had some constructions not known in Britain, such as its impressive stone rows. Thom's work in Carnac was published as *Megalithic Remains in Britain and Brittany* in 1978. At that time the age estimate of Carnac's monuments was off by at least one thousand years, and their true antiquity not yet known.

Throughout his work at Carnac, Thom was testing for his unit of measure called the megalithic yard. On-site research by my brother and I in 2010 at Le Manio showed this unit originated there through making a three-year day-inch count. One imagines the megalith builders adopting this measure as being truly objective since it was created by the invariant relationship between sun and moon over their respective years (and defined by the four-square rectangle). In retrospect, their decision would prove inspired as it enabled the construction of circular simulators (see "Re-creating the Cosmic Clockwork" later in this chapter) and circumpolar observatories (chapter 4).

The unique kerb structure of Le Manio's Quadrilateral and its outlying menhir were built on a significant hillock at Le Manio, just west of the start of the Kermario alignments and visible from the other stone rows to the west-southwest. This kerb structure has four sides not at right angles to each other. Its walls, rarely rising above thigh height, are near-contiguous butted stones with notable large gaps to the west (see figures 3.2 and 3.3). When surveyed the southern and eastern kerbs of the Quadrilateral prove crucial in having provided distinct day-inch counted lengths between their features.

On the southern kerb of the Quadrilateral, two large toothlike stones form a "sun gate." From the gap between the sun gate's stones (point P in figures 3.3 and 3.4) both summer and winter solstice sunrises can be viewed, looking to either stone R on the eastern kerb or to the Menhir Geant, respectively. These two alignments to the solstice sunrises follow a four-by-three rectangle's diagonal angle relative to east, as two 3-4-5 triangles (with a common side lying east-west). The carved edge of stone R, which aligns with the summer solstice sunrise

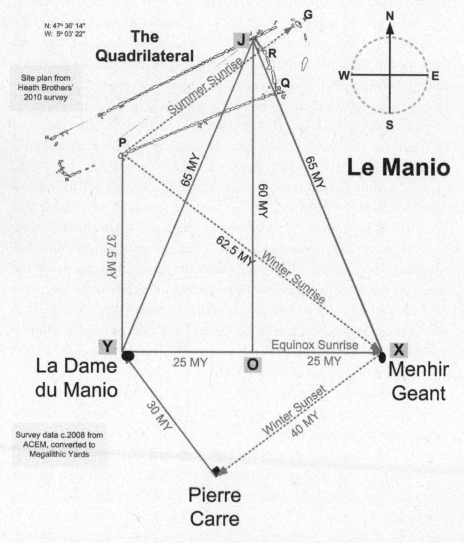

Figure 3.3. The layout of the Le Manio site based upon two recent surveys by ACEM (Association pour la Connaissance et l'Etude des Megalithes) and Heath brothers.

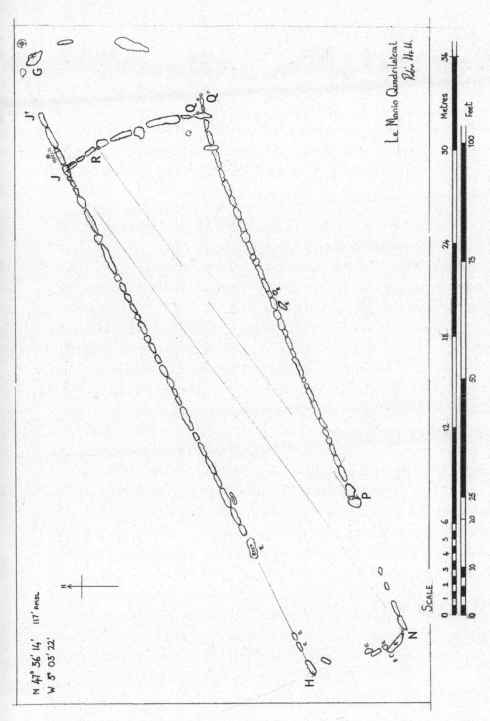

N 47° 36′ 14′
W 5° 05′ 22′
117′ AMSL

N

SCALE

0 1 2 3 4 5 6
10
20 25
50
75
100

Metres 36 30 24 18 12
Feet

Le Manio Quadrilateral
Robin Heath

Figure 3.4. The Heath Brothers' 2009 survey of the Quadrilateral, more reliable than Thom's 1978 plan, of unknown provenance. Drawing by Robin Heath.

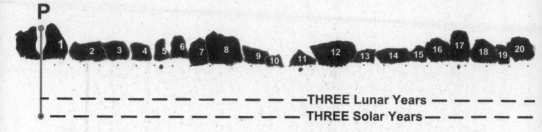

Figure 3.5. Silhouette of the Quadrilateral's southern kerb in which the number of stones, 36 and 37, equates lunar months with the exact day-inch lengths for three lunar and three

as viewed from the sun gate, is 1095¾ day-inches (exactly three solar years) from the sun gate, evidently a specifically chosen distance during the site's design. Beyond that point lies a large stone G with a groove cut in its top edge to further show the alignment to the summer solstice sunrise from point P (the sun gate).

The angle from the sun gate to stone R on the eastern kerb is 14 degrees,* the invariant or "signature" angle of a four-square rectangle's diagonal. The distance between the sun gate (point P) and stone R was three solar years in day-inches, and the ratio between this length and the base of the four-square rectangle (about 36:37) meant the southern kerb, from point P to point Q, was built to represent three lunar years, 36 lunar months, to stone 36. Point Q is found two-thirds of the way along stone 36 of the southern kerb. Point Q', at the eastern edge of stone 37, is the same distance from point P (within the sun gate) as the carved edge of stone R, both these points then marking the end of a three-solar-year count. The end of stone 37 probably aligned by the megalithic builders when a rope of 1095¾ day-inches (three solar years) was brought down from stone R to point Q'. The distance Q to Q' (the difference between three lunar and solar years) is the megalithic yard of 32⅝ inches. This probably sacred unit, derived by the ancients from the sun and moon, was then used to lay out the rest of Le Manio's features (see figure 3.3 on p. 46), which all lie at the corners of four-by-three

---

*This became evident during the summer solstice of 2009 when my brother Robin Heath was looking from point P with a theodolite. He had sought such a clear display of this angle in Britain but never found it.

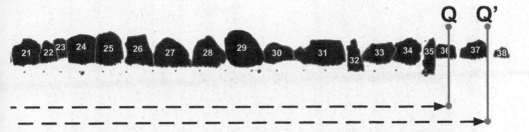

solar years. The distance between Q and Q' is the length of the megalithic yard, measured to be 32⅝ day-inches. From survey photos within the structure and reversed for consistency.

rectangles, each defined in whole numbers of megalithic yards and, through their geometry, capturing the extreme alignments of solstice sunrise in summer and winter.

It looks as though the builders of the Quadrilateral conducted a day-inch count toward stone G over three years, while also using stones to symbolize the passing of lunar months within the count when building the stone monument to commemorate their result. As there were no written records, one should probably view such a monument as fulfilling the need to clearly record the meaning of their enterprise, functioning as a kind of "textbook in stone." A system of cues would help the megalithic student grasp the monument's meaning—here a crucial early result in day-inch counting. Within this monument one therefore finds delightful use is made of the shapes of stones. Some stones, like stone R, were obviously sculpted to communicate *exactly* where the three-solar-year count ended. That the count ran along the summer solstice sunrise alignment to "grooved stone" G was another such cue.

It appears the builders went further and interpreted this monument using an engraved stone, which, like our information boards today, gave extra visual aids and references through which the monument could be better understood. A sophisticated and intuitive geometrical language was developed, involving engraved lines on a set of such stones, now only to be found within the chambered cairn called Gavrinis, six kilometers east of Le Manio. While one finds engraved "outliers" in British monuments such as Long Meg in Cumbria and similar designs in Irish chambered "tombs," nothing can compare with the unique metrological graphics of Gavrinis, whose stones

Figure 3.6. The stone L6 (sixth stone, on the left wall) in the passageway of Gavrinis, which presents the primary lesson of Le Manio's Quadrilateral.
Courtesy of Laurent Lescop.

were gathered from many sites, toward the end of Carnac's formative period, around 3500 BCE.

Twenty-eight finely engraved stones were preserved in the Gavrinis chambered cairn, and one of these apparently came from Le Manio. The stone was moved and reset, with twenty-seven other fully engraved stones, to form part of a passageway and chamber. It became the sixth

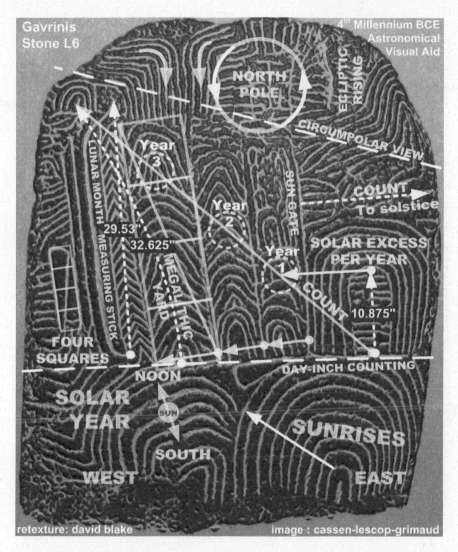

Figure 3.7. Interpretation of stone L6 of Gavrinis as an exposition on the three-year count. Time moves from right to left within the central belt of the stone. On the right the excess of a single solar year over a lunar year (10.875 day-inches) is declared as the outline of a stone having ten horizontal engravings. The three mounting columns then show the growth of this excess over three years. On the left a bow-shaped measuring stick is shown, which references the 29.53 day-inches of a lunar month between its right hand tips. The base of the third year shows a vertical reference mark similar to that on the top left, above the bow, and these reference the megalithic yard of 32⅝ day-inches, again as an exact length.

stone on the left of the Gavrinis passageway and hence was recently called stone L6 in a 3D survey by a team from the University of Nantes. When I was informally invited by Laurent Lescop to view and interpret their work, stone L6 became comprehensible because a number of exact metrological lengths within its engraving corresponded to Le Manio's Quadrilateral.

As shown in figure 3.7 (see p. 51), the lunar month of 29.53 day-inches and megalithic yard of 32⅝ day-inches are both to be found within this engraved stone, as well as three vertical columns representing 12 months, 12 months, and 13 months, respectively. Also shown is a rounded megalith shape, equal to one third of the megalithic yard, that is, the 10 and ⅞ day-inch lunar excess within each solar year. On each of the three columns, hillocks of horseshoe shapes appear to represent the growth of this excess over a three-solar-year count.

The lunar month of 29.53 day-inches was presented by a "bow," which may be a representation of an actual megalithic measuring stick. The megalithic yard is given as 32⅝ day-inches between the base of year three and the reference line above the measuring bow.

The bottom skirt of the stone appears decorated to represent the world of horizon events, as a tunnel of counted units such as day-inches or quite probably months. There are complex markings at the top of the stone that appear to reference the developing world of their astronomical understanding. This division into three horizontal belts is quite common in the engraved stones, as is the grooved banding style indicative of time factoring in units of around an inch.

No engravings would have survived if such stones not been taken into Gavrinis. Erosion, reuse, disregard, and religious vandalism have eliminated such stones when not protected. There are engravings in other dolmen that have not fared nearly as well. Once eroded or vandalized, they become much harder to interpret as references to anything concrete, such as an astronomical invariance, units of measure, or phenomena such as eclipses.

Returning to the Quadrilateral, as revealed by our reconstructed knowledge of day-inch counting and multiple-squares, four types of year related by the three-square and four-square rectangles can be seen represented by the portion east of the sun gate, as shown in figure 3.8.

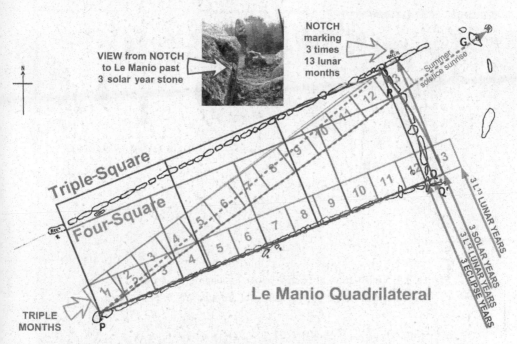

Figure 3.8. The Quadrilateral's shape can be interpreted as a study in four kinds of year: eclipse year, lunar year, solar year, and thirteen-month year, using a thirteen-square bridge between the four-square and three-square. Relative to the southern kerb the northern kerb drops by the diagonal angle of a thirteen-square and then, relative to east, is at the angle of the diagonal of a double-square and hence points to the northern minimum standstill of the moon from Le Menec.

THE TIME TEMPLE OF LOCMARIAQUER

The Locmariaquer peninsula is located southeast of Le Manio and directly west of Gavrinis Island and its chambered cairn. The megalithic sites of Locmariaquer are made up of four or five satellite sites, forming a little known geometrical pattern that resembles, for good reasons, the design of Le Menec's western cromlech (see figure 2.16 on p. 39). These sites were all organized around a singular point called Er Grah where large menhirs were raised, including the largest sculptured menhir ever created in Europe, which now lies broken.

The plan for Locmariaquer was based upon the triple-square, but aligned to north rather than the eastern or western horizon where solar and lunar events occur. The use of north represents the circumpolar

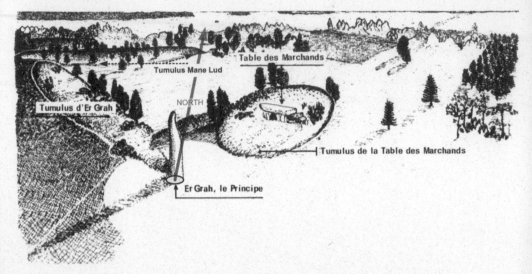

Figure 3.9. Visualization of Locmariaquer looking north. From Association Archeologique Kergal (AAK), *Etudes et Travaux,* vol. 6, p. 15, fig. 5.

Figure 3.10. Refurbished Table des Marchands with broken Grand Menhir Brise beyond. The Grand Menhir Brise was a massive sculptured stone, some of whose parts were used as capstones for dolmen at Table des Marchands, to the right, and the chambered cairn of Gavrinis, 4 kilometers directly east of Er Grah. Photo courtesy of Jim Appleton.

stars, which increases the scope of megalithic astronomy beyond the simple horizon astronomy currently expected (see chapter 4).

The main components of the Locmariaquer design were:

Er Grah (E.G.): a socket at the southern focus of the site, apparently too small for its broken menhir (the Grand Menhir Brise) whose

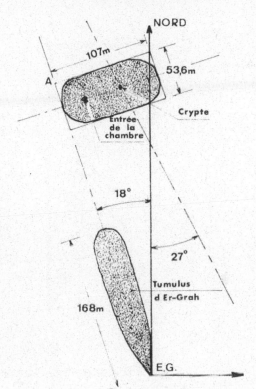

Figure 3.11. Er Grah (E.G.) and Er Grah tumulus (left) with Mane Lud tumulus at top. From AAK, *Etudes et Travaux*, vol. 7, fig. 27.

pieces lead away to the east. One missing piece has been found as the capstone of the chamber of Gavrinis, evidently linking Locmariaquer with the meaning of Gavrinis. Menhirs at Er Grah could be viewed from many other sites and across the bay of Quiberon, then forming alignments with the extremes of sun and moon, according to the multiple-square geometries.

Er Grah Tumulus: a unique swath of rocks, forming a low but long tumulus traveling 18 degrees west of north. This tumulus was longer before recent gardens encroached upon its northern extent, and its length is said to have been 168 meters. Nineteen eclipse years (the Saros eclipse cycle, the period of 18 solar years and 10 days between similar eclipses) would be 167¼ meters, just ¾ of a meter or 29.53 inches less than the tumulus's estimated length, the same as the day-inch count for a lunar month! A triple-square laid along the length of this tumulus defines, through its diagonal lying directly north of Er Grah, the dominant periodicity for the repetition of the sun, moon's phase, and their

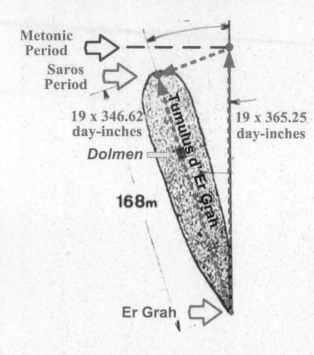

Metonic Period

Saros Period

19 x 346.62 day-inches

19 x 365.25 day-inches

Tumulus d'Er Grah

Dolmen

168m

Er Grah

Figure 3.12. The Tumulus of Er Grah, before it was encroached to the north, was shown rounded and would have given a day-inch length for the Saros eclipse period and the Metonic period.

location against the zodiacal stars—the Metonic period of 19 solar years (see figure 3.12).

Er Grah Dolmen: Within the Er Grah tumulus is a dolmen. The distance in day-inches along the tumulus from Er Grah to the dolmen is 12 lunar years of 13 lunar months. If you draw a line from the dolmen to the line extending directly north out of Er Grah, the distance from their crossing point to Er Grah is 12 solar years. The dolmen is west of that northern point by exactly 4 solar years or 1,461 day-inches, the number of days between leap years. This length is also found in the diagonal of Le Manio's Quadrilateral. Forming a right angle with the same triple-square diagonal along the tumulus from the 12 solar year point on the north line from Er Grah reveals the length of 12 eclipse years. The difference between 12 solar years and 12 eclipse years is 18.618 feet (18 feet, 7½ inches), the same number of days taken for the lunar node to move the distance the sun moves in angle each day.

Mane Lud Tumulus: Here we again see the triple-square angle of 18 degrees, this time 18 degrees north of east so as to form a right angle with the 18 degrees west of north line formed by the Er Grah tumulus. The Mane Lud tumulus forms the shape of a rounded double-square.

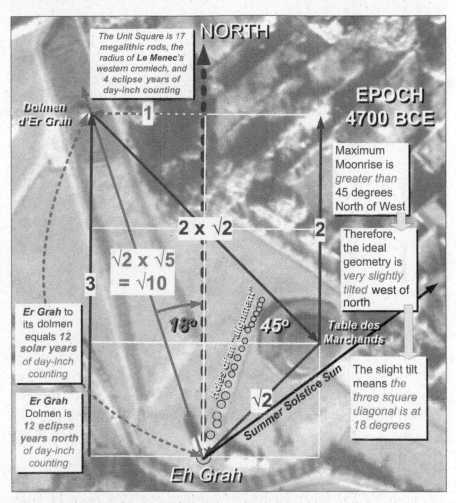

The Unit Square is 17 megalithic rods, the radius of **Le Menec**'s western cromlech, and 4 eclipse years of day-inch counting

NORTH

Dolmen d'Er Grah

1

EPOCH 4700 BCE

Maximum Moonrise is *greater than* 45 degrees North of West

2 x √2

2

Therefore, the ideal geometry is *very slightly tilted* west of north

√2 x √5 = √10

3

18°

45°

Table des Marchands

Er Grah to its dolmen equals *12 solar years* of day-inch counting

holes of an "alignment"

The slight tilt means *the three square diagonal is at 18 degrees*

Er Grah Dolmen is *12 eclipse years north* of day-inch counting

√2

Summer Solstice Sun

Er Grah

Figure 3.13. The geometric alignment and metrology between Er Grah, its dolmen, Table des Marchands, and the radius length of Le Menec's western cromlech, including the triple-square with diagonal of 12 solar years from Er Grah to its dolmen. The dolmen would have been aligned to Ursa Major tail star Alkaid from Er Grah.

Its eastern end touches the line that comes directly north from Er Grah at a distance of 33 solar years of day-inch counting. The tumulus's 18 degree angle north of east places the *western* end of the tumulus at a distance of 33 eclipse years from Er Grah (see figure 3.20 on p. 63).

Table des Marchands: This impressive dolmen within a circular tumulus has another piece of Er Grah's sculpted menhir as its capstone. The stone presents a rotating axe motif that, seen from below, appears

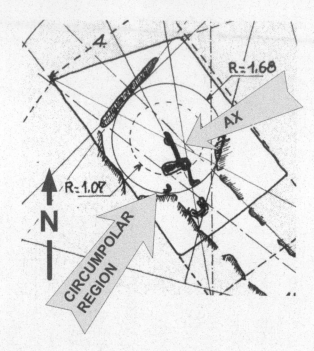

Figure 3.14. The site plan of Table des Marchands looking through roof stone at axe below. The axe represents the northern orbit of a circumpolar star (see also fig. 5.1). Adapted from AAK, *Etudes et Travaux,* vol. 6, p. 41, plate 6.

to model the rotation of a circumpolar star counterclockwise. When part of the Grand Menhir, this capstone was contiguous with the roof stone of Gavrinis, but as it is set now its carvings point downward. The star represented by the axe appears to be the tail of Ursa Major's "handle," called Alkaid.*

Further north of Er Grah is the original tip of its tumulus, which was probably 223 lunar months of day-inch counting from Er Grah originally. This tip would have represented the 223 lunar months of the Saros eclipse cycle, the dominant eclipse cycle predicting very similar eclipses over its periodicity. A line at right angles to the tumulus would pass the meridian north of Er Grah at a distance of 19 solar years, the length of the Metonic period.

Further north again the eastern tip of the tumulus of Mane Lud is about 12,050 day-inches north of Er Grah or 33 years of day-inch counting. The tumulus of Mane Lud was obviously intended to terminate the Er Grah complex to the north, being inclined to form a right

---

Alkaid is Arabic for "leader of the mourners," Ursa Major being seen as a funeral bier to carry the dead "king." Ursa Major is also associated with the bear, *Arktos* in Greek.

Figure 3.15. The three main components of the immediate monument of Er Grah looking east: (I) some pieces of the broken Grand Menhir Brisé, (2) the Er Grah tumulus, which stretches 18 degrees west of north from Er Grah, and (3) Table des Marchands, whose capstone was originally part of the Grand Menhir. Photo from Er Grah visitor's center notice board.

angle with the Er Grah tumulus radiant, which lies 18 degrees west of north from the Er Grah point.

Locmariaquer's Er Grah complex is therefore a highly structured map of time, expressing time scales that define geocentric time on Earth. The dolmen of Er Grah establishes its purpose, at the diagonal angle of a triple-square, and 12 times 13 lunar months, that is 12 years of 13 months *or* 13 years of 12 months. Then the Er Grah tumulus establishes the crucial framework between the Saros and Metonic periods, while the 33-year scope of Mane Lud in the north provides the period of the Solar Hero in myths (heros like Jesus, Krishna, or Mithras dying when 33 years old). The sunrise will return to exactly the same spot on the horizon after 33 years because the fractional part of the solar year is almost exactly 32/132 days and 132 equals 4 times 33.

This brings us back to the dolmen of Er Grah, which reveals the difference between 12 solar years and 12 eclipse years, 242 days, which in day-inches equals 18.6154 feet, the number of *days* taken by the lunar nodes to travel the distance *in angle* traveled by the sun each day. That is, the Er Grah dolmen was probably placed, in time, where it had become easiest to measure this key nodal rate in feet.[*]

---

[*]The complete nodal period is 18.618 years exactly because it takes 18.618 days to move 1/365.2422 of the ecliptic, so if 18.618 days of nodal motion equal one day of solar motion, then 18.618 years of nodal motion equal one year of solar motion. The solar year of 365.2422 days divides by the nodal motion of 18.618 days 19.618 times so as to form an N:N+1 relationship between the solar year, the eclipse nodes, and the solar day.

Figure 3.16. The Dolmen of Mane Lud. Photo courtesy of Vincent Lefèvre, http://locker56740.free.fr.

The Dolmen of Mane Lud (Key to Locmariaquer)

The tumulus of Mane Lud is very large and originally was even larger. It is very likely that the tumulus was added to preserve the nature of its monuments: the dolmen (formerly central to the western half of the tumulus) and a "triple mountain" tomb, symbolically drawn by a team of horses.

The original tumulus was modeled upon the double-square but angled 18 degrees north of east to complement the triple-square angle of Er Grah's tumulus 18 degrees west of north. The two complementary angles together achieve a 90-degree right angle and point to the diagonal angle of the single-square, that is the 45 degrees of the extreme northern moonrise at maximum standstill. The eastern half of the tumulus involved a dome-like structure containing a sophisticated "tomb" for two halflings, the right halves only, one blackened and the other white. They held reins to a double arc of posts, some topped by horse heads. This apparently represented the halflings as riding the moon's extremes.

It would have been useful for periods in day-inches, when multiplied

Figure 3.17. In 1864, the tumulus to the west was already eroded to the dolmen and a large subsurface cairn and cist were discovered. Adapted from AAK, *Etudes et Travaux*, vol. 7, p. 12, fig. 3.

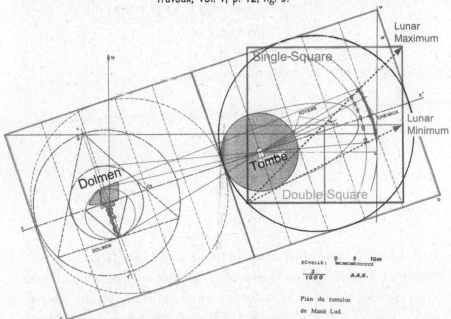

Figure 3.18. The design of the Mane Lud tumulus had two parts defined by a double-square and at a triple-square angle. The dolmen, 1/40th scale model of the greater complex, points down to Er Grah. The eastern half was modeled on the horse-drawn chariot, with two halflings riding the moon's extremes, naturally associated with the single-, double-, and triple-square geometries at Carnac. Adapted from AAK, *Etudes et Travaux*, vol. 7, part of centerfold.

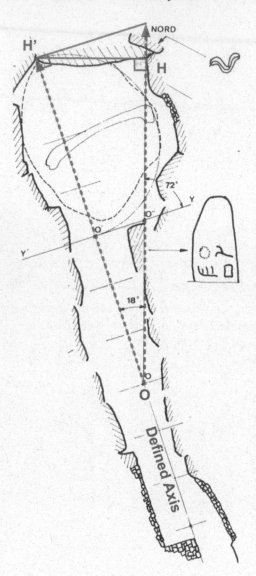

Figure 3.19. Internal plan of the Mane Lud dolmen showing its axis mimicking the Er Grah axis from Er Grah (point O) to Mane Lud western edge (point H') and Mane Lud southeastern corner (point H). Adapted from AAK, *Etudes et Travaux*, vol. 7, p. 42, fig. 24.

by twelve, to yield their same numerocity in the larger unit of feet because when things stay the same in number but the units of measure are a different size, a change of *scale* automatically results. Our plans and maps are almost always scaled down, and when there is doubt as to how the megalithic people could have worked on a project as big as Locmariaquer, the dolmen shows that a scale model of the whole monumental complex existed where the dolmen now is. More than 1,000 feet were reduced to just over 25 feet, and the axis for the corridor was set

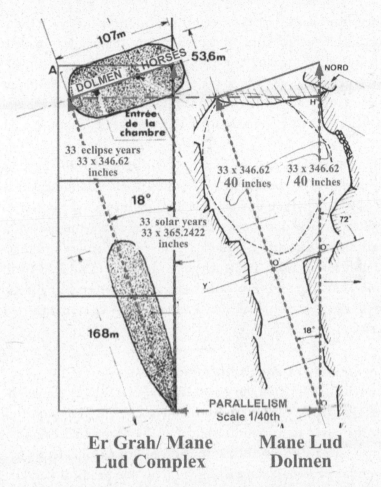

Figure 3.20. Mane Lud dolmen demonstrates that a scale model was built either to direct its building or to explain its organization, or both, using a scaling factor of 1/40.

to the triple-square angle of the Er Grah tumulus from a specific reference point, representing Er Grah (see figure 3.19, Er Grah represented by point O).

Because the Locmariaquer complex was to have a northern dimension of 33 solar years, the builders used a unit of 33 inches as this would divide into the monument 365.2422 times, as days divide the solar year. A yard of 33 inches* is then 11/12 of the familiar English yard and, like

---

*This unit survived well into history as, for example, Akbar's yard and the Spanish unit, the *vara*. It is also related to the Saxon foot of 1.1 feet.

the megalithic yard (just three-eighths of an inch shorter), it contained
40 megalithic inches. Using this property the dolmen was made (probably before anything else) as a 1/40 scale model of Locmariaquer from
Er Grah to Mane Lud. The shape of the Mane Lud dolmen monument
mimics that of the greater monument, and the length of the dolmen,
from a clear starting point, is 365.2422 megalithic inches of a 33-inch
yard.

By employing a unit of measure that divided into the Locmariaquer complex, the builders could represent the monument using an
established subdivision of that unit. Scaling invokes the power we call
multiplication and division without resorting to a symbolic arithmetic
unavailable to megalithic builders. The solar year was originally the
super unit for a count of 33 such years. By then using a 33-inch unit,
the solar year count was exposed as a *measurement*, enabling megalithic
inches in the dolmen to scale down any aspect of Locmariaquer's design
without doing any arithmetic.

The Solar Hero's Coach and Horses

The seventh volume of the *Association Archeologique Kergal* (AAK)
Etudes et Travaux, "Le Tumulus de Mane Lud . . . Montagne Solaire,"
has on its cover an astounding sight of the Mane Lud tumulus with
two men, one white, one black, in the "cab" of a stone dome, upon an
oval terrace of stones, controlling via reins the heads of horses set on
posts.

This extraordinary construction, now beneath the Mane Lud tumulus, was built in the eastern square of the tumulus and only the right
halves of the two men were interred into the tomb. These were laid
diagonally across the tomb and connected to the horses through reins.
Figure 3.22 on p. 66 shows the highly geometric layout due to multiple
squares and inscribed circles, found by the AAK.

The function of this part of Locmariaquer, positioned at the top
of 33 solar years of day-inch counting from Er Grah, suggests that the
moon's nodal cycle was seen as relating to the 33-year cycle. Figure
3.22 gives a close up of the arrangement found during excavation. One
can see that the posts form two sections of two circles, both concentric on the tomb. These arcs, by their staggered termination relative

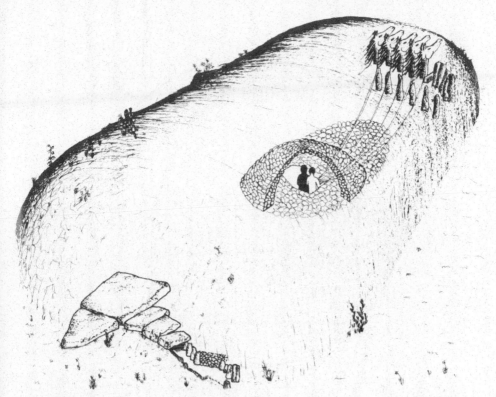

Figure 3.21. Cover art of *Association Archeologique Kergal Etudes et Travaux*, vol. 7, "The Tumulus of Mane Lud . . . Solar Mountain," showing dolmen (bottom left), White and Black halflings (center), and posts, some topped by horse-head skulls (top right).

to one another, define two lines that cross on the inscribed circle of the tumulus' eastern square. The angle then formed between the two rows of posts at that point defines the range of the moon's maximum and minimum standstills, as shown by the single- and double-squares in the figure that combine with the underlying triple-square angle of the tumulus. Seen in this way, one notices that only the outer posts and, of these, only those to the north of the midpoint (i.e., the sun's solstitial angle) were surmounted with horse heads. This suggests that the moon was seen to be pulling on the solar chariot when it exceeded the solstice angle, during the maximum half cycle of the lunar nodes.

The white and black right-handed halflings seem to symbolize

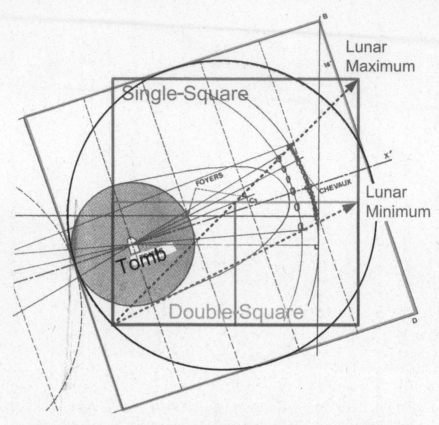

Figure 3.22. The eastern square of Mane Lud showing how the multiple squares resolve in the staggered poles of two concentric circles, demonstrating the moon's excess over the sun for half of the nodal cycle. Adapted from AAK, *Etudes et Travaux*, vol. 7, part of centerfold.

the half of the Earth defined by the local meridian, which is always travelling east as the earth rotates. The meridian is therefore *the right half of the Earth* from north to south—the local meridian, which is half white by day and half black by night. With their heads to the north, they were eternally traveling east. This could explain the viewpoint of the tomb and of the two half-men as directly symbolizing an astronomical situation rather than their being "sacrificed" for obscure reasons. The reins to the horse-headed posts probably represented a megalithic astronomical concept built as a profound "installation" and subsequently preserved by placing the tumulus over both it and the dolmen.

Figure 3.23. The dolmen of Mane Ruthual. Photo courtesy of Vincent Lefèvre,
http://locker56740.free.fr.

MANE RUTHUAL AND THE ECLIPTIC POLE

To the southeast of Er Grah and forming a 3 (east) by 4 (south) rectangle is Mane Ruthual, a large whale-shaped dolmen whose chamber is
aligned to the lunar minimum moonset in the north. Mane Ruthual
is 254 meters or 10,000 inches to the south (the meter being equal to
10,000 divided by 254, or 39.37, inches). Since there are 254 lunar orbits
in the Metonic period of 19 years, the monument appears naturally
related to the Metonic period, but now in lunar orbits per meter and
not in day-inches. Er Grah, seen from Mane Ruthual, would also have
the ecliptic pole sitting at its maximum western elongation, directly
above Er Grah, once per day when the solstice extreme of the ecliptic
was rising.

It might seem that an alignment to the ecliptic pole would be too
advanced for the megalithic to grasp, being—like the north pole in that

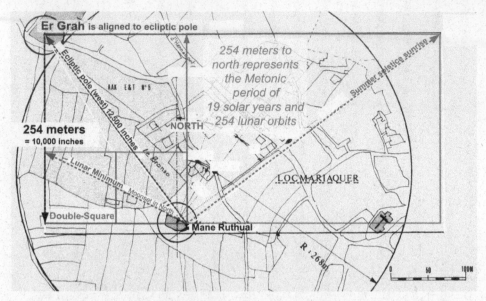

Figure 3.24. The significant location of Mane Ruthual relative to Er Grah as part of a wider pattern of monuments at Locmariaquer.

era—just an empty location in the night sky. But as the methods of circumpolar astronomy come into focus, it becomes clear that the ecliptic pole is less understood today than it was by ancient circumpolar observers. The ecliptic pole had a natural relationship to solstice sunrise and sunset when viewed near Carnac. There it was aligned, as an azimuth *to the north,* through the 3-4-5 triangle's smaller angle, as shown in figure 3.24. This pattern of monuments at Locmariaquer resembles the example given of the canevas in chapter 2, a geometry clearly organized in a cross, angled west by the smaller angle of a 3-4-5 triangle (as is shown by figure 3.25). This pattern is aligned to the ecliptic pole in the northwest and the summer solstice sunrise in the east.

This same alignment is to be found at Le Menec's western cromlech, which incidentally has a perimeter equal in length to the 10,000 inches or 254 meters displacement south of Mane Ruthual from Er Grah. The corridor leading to the chamber appears based upon a four-square with its diagonal aligned east-west (figure 3.26). The chamber was then set to run at the diagonal angle of the double-square, so as to have its point toward the moon's minimum moonset to the north. The far edge of the capstone aligns to north (as does Mane Lud's capstone, but then

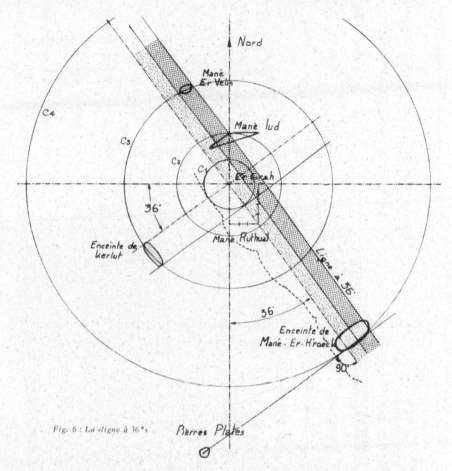

Fig. 6 : La ligne à 36°.

Figure 3.25. The signature angle of Locmariaquer's monuments aligns with the angle of the 3-4-5 triangle. Adapted from AAK, *Etudes et Travaux*, vol. 5, p. 16, fig. 6.

from the west), emphasizing the single-square alignment to moonset at lunar maximum in the north. The ecliptic pole alignment to Er Grah is a further double-square that on top of Mane Ruthual's double-square, achieves the larger angle of a 3-4-5 triangle.

If it were a piece of jewelry, the geometry of Mane Ruthual would seem a magnificent and appropriate setting for what lies in its chamber. It is a double chamber and hence congruent with its double-square alignment. In the first chamber one stands in a similar floor plan as the carving, still just visible, on the ceiling of the second chamber. The AAK tried hard to obtain a metrologically accurate plan for the ceiling design, as in figure 3.27.

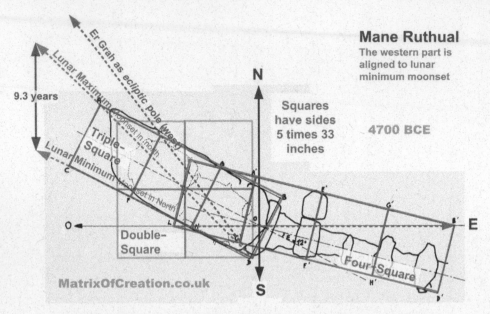

Mane Ruthual
The western part is aligned to lunar minimum moonset

N

Squares have sides 5 times 33 inches

4700 BCE

9.3 years

Er Grah as ecliptic pole (West)

Lunar Maximum moonset in north

Triple-Square

Lunar Minimum moonset in North

O ← → E

Double-Square

Four-Square

MatrixOfCreation.co.uk

S

Figure 3.26. The geometric elements that define the dolmen of Mane Ruthual. Adapted from AAK, *Etudes et Travaux,* no. 5, page 76, figure 44.

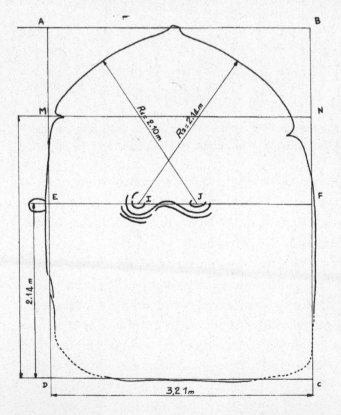

A   B

M   N

R₁ = 2.10m   R₂ = 2.14m

E   I   J   F

2.14 m

D   3,21m   C

Figure 3.27. The plan of Mane Ruthual's carved ceiling showing two distinct centers in an apparently crude attempt at an onion-shaped form, a widespread architectural form for later civilizations. From AAK, *Etudes et Travaux,* no. 5, page 71, figure 40.

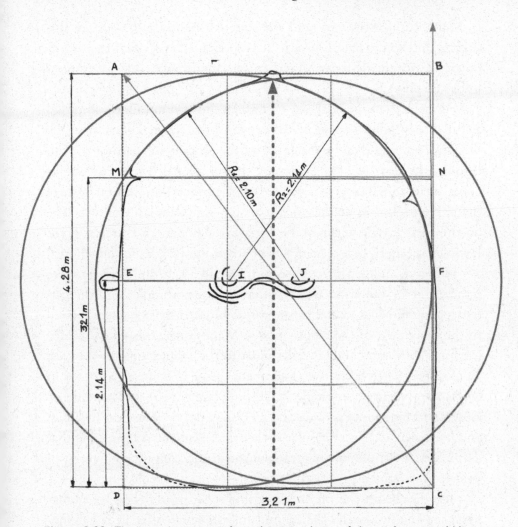

Figure 3.28. The invited expansion from the given design of the roof carving of Mane Ruthual. The asymmetry of the radial arcs transforms into the Earth's celestial equator, center I, and the sun's path or ecliptic, center J, which travels over the menhir at point E, placed below the extreme winter solstice sun in the west. The dimple at the top is the point of crossing for the sun over the celestial equator at equinox. Adapted from AAK, *Etudes et Travaux*, no. 5, page 71, figure 40.

This design is based upon two foci that each act as the center for one of two non-concentric circles. It was therefore a very early expression of double circles that associated with the later generation of domes, arches, and, in the simplest form, the *vesica pisces* in which points I and J would each sit one radius apart and hence upon each other's perimeter.

In Mane Ruthual the two radii are centered slightly apart but are also slightly different in length so as to achieve a radial arc from J that *intentionally* leaves the figure at M, rather than inscribing a full quadrant as I does, to point F.

Figure 3.28 on p. 71 shows that when the two radii are completed as full circles, point I could be the center of the celestial equator in the sky while J would then be the center of the ecliptic, the path of the sun through the year. The two circles are the traditional "world parents" of many ancient mythologies, separated by the process of creation. Where the two circles cross at the dimple on top would be an equinox. The north pole is at right angles to the celestial equator and is therefore point I, while the ecliptic pole is point J, similarly at right angles to the ecliptic.*

However, another interpretation is equally likely and perhaps more indicated. One can argue that the Locmariaquer complex and the pattern of monuments near Carnac are dominated by behaviors related to the movement of the moon's nodes. Mane Ruthual is tilted toward the lunar minimum and any alignment to the ecliptic pole in the west "comes free" when there is an alignment to the midsummer sunrise in the east and plenty of multiple squares implicit within the monument's structure. Mane Lud expresses the moon's nodal cycle within its coach and horsemen and it is the solar cycle contra tent the eclipse cycle that dominates the triple-square arrangement of successively long cycles of twelve, nineteen, and thirty-three years.

The dimple on the tip of the ceiling relief would then be a lunar node and the two circles the ecliptic and the lunar orbit. If the monolith to the left is the moon standing low in the northwest, then point J is the center of the moon's orbit. But behind these possibilities comes a realization that change comes out of cyclicity, tilted with respect to each other—the Earth's axis angle to the ecliptic giving us the seasons of the year and the precession of the equinoxes and the lunar orbit tilted to the ecliptic giving the extra variation of the nodal year, adding to the varia-

---

*In retrospect it is possible to find stones at Le Manio as foresights to the ecliptic pole in both its eastern or western elongations. Also, Le Menec's western cromlech has the eastern elongation marked by an outlier from the early western alignments, while its major axis is, as already mentioned, the western elongation alignment, both of these being relative to the central mid-sight for the cromlech's observatory.

tions between the moon and sun, seen in the triple- and four-square relationships.

The principle expressed, in either case, means that the extensive nature of time is, through these circles in the sky, tilted with respect to the Earth's axis, and this tilt introduces an extraordinary numerical pattern, a kind of intensive structuring of time.

One or other of the above interpretations would show this image as an excellently presented piece of visualization and a wonder in human understanding. However, it is debatable whether one should call the Mane Ruthual roof art a picture of the Goddess or whether it was simply a diagram showing a hard won piece of astronomical knowledge. Projecting religious meanings onto such a technical diagram must first establish how our religious symbolism came into existence and whether astronomical knowledge was indeed one of the sources inspiring it.

RE-CREATING THE COSMIC CLOCKWORK

As far as we can tell, Le Manio's picture of time was restricted to about four solar years, the length of 1,461 day-inches across its diagonal. This appears to have related to a significant periodicity for eclipses, today called the Octon of 4 eclipse years, approximately 47 lunar months. The Octon period makes it highly likely for there to be some sort of lunar eclipse 47 lunar months after a previously observed eclipse. This eclipse period is far less accurate than the Saros period of 18 years (and ten days), which provides eclipse repetitions so similar that today each eclipse is given a number in the Saros Series. However, when working over 3 to 4 solar years, the three solar years gives a weak but useful anniversary for the sun and moon of 37 lunar months while 4 eclipse years gives a weak but useful anniversary for eclipses over 47 lunar months. The scope of the Quadrilateral is therefore an early echo of the Saros and Metonic periods of 19 eclipse years and 19 solar years. Indeed, there are 235 lunar months in the Metonic period, which is 5 times 47 months. Locmariaquer not only completed the visual layout of time through the 19 years of the Metonic period, including the Saros period, but also added the 33-year period while developing the theme of the triple-square's relationship between the solar year, eclipse year, and 13-month year.

To measure time more accurately than by counting days, new instruments were needed. The first was an instrument that would allow the movement of the sun and moon to be tracked directly upon the ecliptic and therefore independently of horizon observations. The second instrument would measure time directly from the motions of the northern or circumpolar stars. This circumpolar area grows larger as one travels north upon the Earth, as more and more stars never set. These stars can provide an accurate measure of time in units of less than a day because their motion is directly due to the rotation of the Earth. Both of these new instruments introduced the geometry of the circle, always suitably calibrated by measuring progress around the perimeter, which we would today call angular motion.

The four solar year length across the Quadrilateral can be turned into a circle without knowing the radius of such a circle. Initially a square can be made of side length one solar year of day-inch counting and the task is to convert this into a circle with the same perimeter. John Michell shows how this could have been achieved through dividing the solar year side length into eleven parts and building in, from the ends of one side, two 4-by-3 rectangles, leaving a single-square between these of side length 3. The center of this three-square is then a point on the circumference and the radius of the circle between that center and the center of the large single-square. The new circle's radius is 7 units of 33.2 day-inches, convenient to the circumference of 44 units, according to the most used pi approximation of 22/7.

Using the new four solar year circle, a sun marker could be moved by 4 inches per day to simulate the sun. The circle then became a model of the ecliptic and could be populated by observing the stars seen at night to build up a picture of the zodiac constellations and brightest stars, organized by later traditions into divisions of 28 lunar mansions or 12 zodiacal constellations. Turning the linear day-inch count into the circle of the sun's path marries the recurrent properties of the circle with the four year inch count, then resembling an *ouroboros,* a serpent that eats its own tail, or, less figuratively, the four year rope allows the sun to return to where it was one year ago. The helical rising of stars, before sunrise, feeds into this elegant simulator of the sun by showing the stars that border the day. It would not have been possible for megalithic astronomers to make such a simulator if they had not been calibrating geometrical constructions using units of length.

A simulator could count independently of time while showing the structure of time in its divisions and overall count. The knowledge of the sun's position could then escape from the limitations of only observing twice daily its alignment on the horizon at rising and setting. The idea that the sun and moon ran around a circular path, with the Earth at the center, replaced the confusing concatenation of earthly rotation and orbital motion with a greatly simplified view of orbital motion lying behind the Earth's daily rotation.

A further step in the development of simulators began with the observation that the moon in the sky sat over the same part of the ecliptic every 82 days. This happens because its orbital period is very close to 27⅓ days. Thus, every third orbit of the moon places the same stars behind the moon as those 82 days before. The question must have arisen as to how 82 would divide into the 1,461-inch circumference of the sun simulator. The result, found by successively laying an 82-inch

Figure 3.29. Gnostic gem from Roman-era Egypt (first century AD), with an ouroboros surrounding a scarab and *voces magicae*, characters representing magic words, often found in early Christian works.

rope over the 1,461 inches of four solar years, is 17, and then a slightly smaller circle could be formed within the solar simulator. A rope 17 times 82 inches long would ensure the new circle would divide into 82 parts, each of 17 inches. These units would be suitable for moving a moon marker by three divisions every day, causing the moon marker to be on the same spot every 82 days.

Parts of exactly this apparatus were discovered at Le Manio in 2010 by my brother and I, during a public meeting involving the creative use of ropes and participants to try adding a Pythagorean 12-13-5 triangle to the southern kerb. Stones were found within the current walkway that were laid lengthwise as if to form part of a circumference of stones whose center would originally have been where stone 29 now stands on the southern kerb.

These simulator stones are 17 inches apart. The circumference is 17 times 82 or 1,394 inches and the radius is then exactly 6.8 megalithic yards. The numerocity of 17 in the simulator can also be detected in 17 being one quarter of 68, while the moon's nodal period is 6,800 days long, making this orbital simulator "belong," through its radius, to the nodal period's signature number of 17, based upon the strange coincidence that the megalithic yard and 82 form a compatible approximation of 2 pi as 820/261.*

Effective simulation of the moon and sun was therefore a natural opportunity and would have been a great step forward for megalithic astronomy, through transforming the 1,461-inch length into a whole number of 82-inch lengths. The moon could now be seen within the context of its journey through the ecliptic's starry backdrop alongside the sun and hence giving warning of impending eclipse conditions. The moon would soon be seen to travel above the ecliptic during half of its orbit and below the ecliptic during the other half of its orbit. This would have exposed the position and also shown the backward or retrograde motion of the lunar nodes through *either* (a) seeing where the moon stood when exactly on the ecliptic *or* (b) seeing

---

*The megalithic yard found in Le Manio's three year count was 261/8 inches long, so pi's close proximity in value to 820/261 allows megalithic yards in a radius to deliver divisibility by 82 in the circumference.

Figure 3.30. The radial stones that could have formed part of a lunar simulator, based upon 82 units equaling the days taken by the moon to orbit three times. Photos taken during Heath's 2010 survey.

where the moon was when standing at its farthest point away from the ecliptic *or both*. It would soon become obvious that the other node was always diametrically opposite upon the circle so the location of the other node could always be inferred. This would have given the astronomers a powerful new framework within which to study the otherwise invisible nodes and to see how these affected the highly variable appearance of the moon.

The remaining problem in using the 82-day recurrence of the moon upon the ecliptic was that horizon astronomy could only see the two lunar events in a day when the moon rose or set. After 82 days the moon would rise at a point on the ecliptic that was *82 days later* in the 365-day solar year and hence the same ratio of a sidereal day *earlier*. The lunar simulator told the astronomers where the moon was on the ecliptic, but they would not yet know which part of the ecliptic was rising when the moon rose.

If the moon was above the ecliptic, it would rise earlier than predicted by the simulator, but if the moon was orbiting below the ecliptic, it would rise later than the part of the ecliptic, shown by the simulator. Without knowing where the sun would have risen, at the ecliptic longitude of the moon, the variation due to the moon's ecliptic, latitude could not be measured.

The location of Le Manio on top of a small hillock probably became too small for the work of quantifying when a specific part of the ecliptic was rising and so the work was re-established at Le Menec, which had a good horizon to the east and a good long stretch of ground upon which to place measurements of the moon's latitude above or below the ecliptic over its nodal cycle—these being what the stone rows of the western alignments represent.

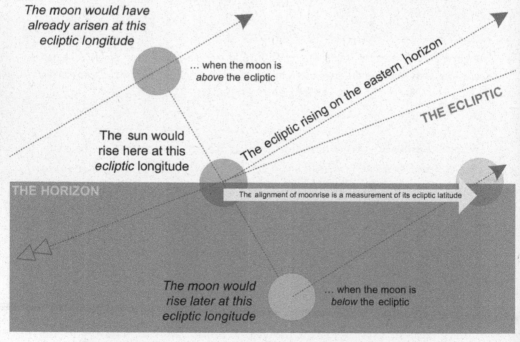

Figure 3.31. The amount by which the moon orbits above or below the sun's path could be measured, using a simulator, once one knew where on the ecliptic the moon was. The moon would rise early or late, but how could one know when a specific point of the ecliptic should rise on a given day of the year?

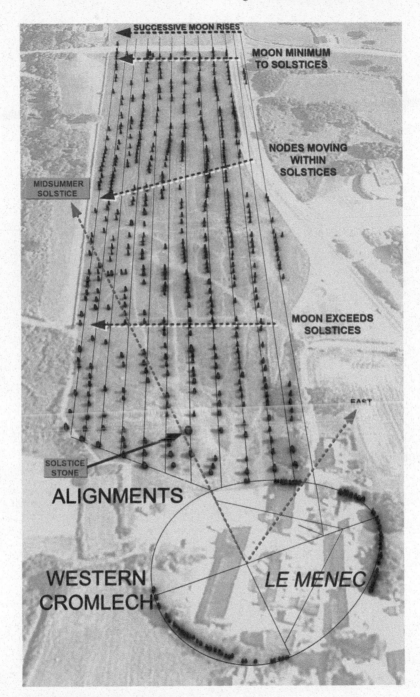

Figure 3.32. Presentation of the moon's early and late moonrise according to its ecliptic latitude in the western alignments of Le Menec. Earth imagery underlay in image provided by DigitalGlobe, 2013. Visualization by David Blake, based upon data by Alexander Thom.

4

THE FRAMEWORK OF CHANGE ON EARTH

THE NORTH POLE is the only fixed point in the night sky. All the "fixed stars" appear to move around the pole as the world rotates. In classical antiquity the lights of the sky were divided into fixed and wandering—the stars were seen as "fixed" and the planets as "wandering." Stars move as if fixed to a black sphere that rotates around the Earth, making it appear that the Earth is at the center of the universe. This geocentric worldview, that the Earth was the center, would not have been what the megalithic people understood judging from the astronomical sophistication of their monuments. By the early medieval period, the thought control of great empires and dumbing down of successive catastrophes had thoroughly eroded the superior worldview demonstrated in Carnac's monuments.

The north pole was crucial to what the megalithic people were able to achieve. Their early technical work, unlocking what the north could reveal, probably led to the symbolic influence of the north found within

Figure 4.1. (Opposite) The interior of Table des Marchands, showing the axe carving on the ceiling stone. Due to the direction of the axe blade and the loop at the base of its handle, it appears that the axe would rotate counterclockwise like the circumpolar stars, when viewed from below. In 1991 the dolmen was refurbished to look like a cairn rather than an open structure.

ancient myths and legends. Stories created as far back as 3000 BCE appear to deal with the Precession of the Equinoxes, an "earth-shaking" phenomenon caused by the pole's slow but regular motion consequently shifting our world's relationship to the stars. In the north, and over thousands of years, the pole describes a circle through the sparse northern stars (see figure 4.2). Using a number of well-crafted metaphors, ancient myths present Great Time as emerging from changes occurring in the sky over thousands of years. The relocation of the spring and autumn equinoxes relative to the zodiacal band of star constellations

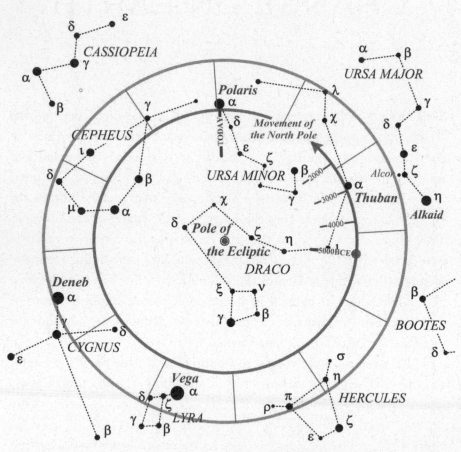

Figure 4.2. The celestial pole is not fixed relative to the stars. It describes a circle in 25,780 years, which millennia ago may have appeared of longer duration. The circle in the figure shows the north pole's position from 5000 BCE to today. Notice how few stars are ever touched by the path of the north pole over its ~26,000-year precessional cycle.

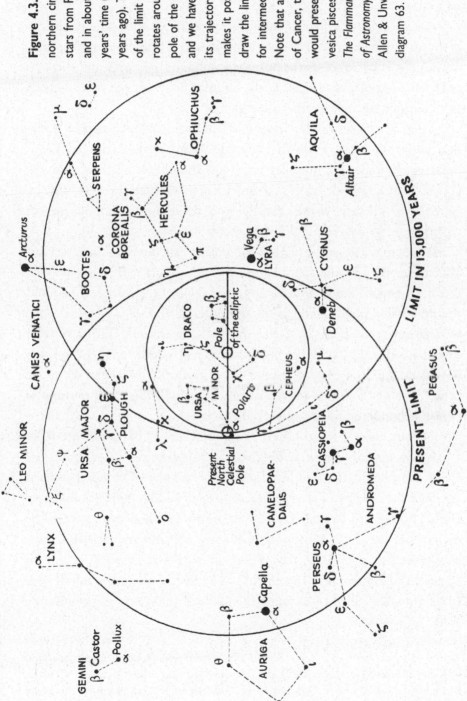

Figure 4.3. Limit of the northern circumpolar stars from Paris, today and in about 13,000 years' time (or 13,000 years ago). The center of the limit circle rotates around the pole of the ecliptic, and we have indicated its trajectory, which makes it possible to draw the limit circle for intermediate dates. Note that at the Tropic of Cancer, the diagram would present a perfect vesica pisces. From *The Flammarion Book of Astronomy* (London: Allen & Unwin, 1964), diagram 63.

was seen as part of a primordial drama, playing out within the life of mankind.

Away from the tropics there is always a circle of the sky whose circumpolar stars never set and that can be used for observational astronomy. As latitude increases the pole gets higher in the north and the disk of the circumpolar region, set at the angular height of the pole, ascends so as to dominate the northern sky at night.

Therefore, the angular height of the pole at any latitude is the same angle we use to define that latitude, and this equals the half angle between the outer circumpolar stars and the pole itself. For example, Carnac has a latitude of 47.5 degrees north so that the pole will be raised by 47.5 degrees above a flat horizon, while the circumpolar region will then be 95 degrees in angular extent.

It is perhaps no accident that the pole is called a pole since to visualize the polar axis one can imagine a physical pole with a star on top, like a toy angel's wand. The circumpolar region is "suspended" around the pole like a plate "held up" by the pole. Therefore, a physical pole, set into the ground, can be used to view the north pole from a suitable distance south (i.e., with the pole's top as a foresight for the observer's backsight). Such an observing pole would probably have been set at the center of a circle drawn on the ground, representing the circumpolar region around the north pole. This arrangement, a *gnomon,** existed throughout history but in recent times presented as part of a sun dial.

It now appears a *gnomic* pole was also used in prehistory to locate the north pole in the middle of circumpolar skies. The north pole is opposite the shadow of the equinoctal sun at midday. The gnomic pole could also be used to find "true north," as located halfway between the extremes of the same circumpolar star above the northern horizon. This can make use of the fact that when the sun is at equinox, it lies on the celestial equator and therefore is at a right angle to the north pole (see figure 4.4). This right angle is expressed at the top of the gnomic

---

*In ancient Greek, *gnomon* means "indicator," "one who discerns," or "that which reveals." According to the testimony of Herodotus, the gnomon was originally an astronomical instrument invented in Mesopotamia and introduced to Greece by Anaximander. I assert that it was innovated even earlier, in the megalithic period, because structures that could operate one still exist within megalithic monuments.

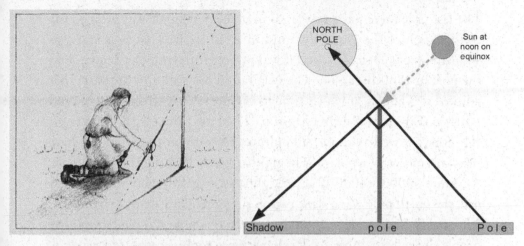

Figure 4.4. From pole to pole. It is possible to determine the angle of the north pole using a gnomic pole as shadow stick, but only at noon on the equinox. The laddie on the left cannot do this to a possible few minutes of a degree, but the geometry of the stick and shadow length can, providing true north and equinox can be determined. Illustration on left from Robin Heath, *Sun, Moon and Stonehenge*, fig. 9.3, by Janet Lloyd Davies.

pole used and hence can enable the alignment of the pole through the similarity (or congruence) of all the right-angled triangles within the arrangement.

To achieve an accurate bearing to true north, a circumpolar observatory can use the gnomic pole method, not just at noon on the equinox but every night, by dividing the angular range of a bright circumpolar star, in azimuth. The north pole's altitude, known to us as latitude on the Earth, can then be identified by dividing the angular range in altitude of a circumpolar star, a task achievable through geometry and metrology, so as to create a metrological model of the latitude upon the Earth.

It would seem obvious today that the pole star Polaris could have been used, but this is a persistent and widespread misunderstanding of the role of pole stars within the ancient and prehistoric world. *Epochs in which there is a star within one degree of the pole are very rare and short-lived.* Our pole star, Polaris (alpha Ursa Minor), is currently placed two-thirds of a degree from the pole that is moving through the northern sky (figure 4.3 on p. 83) in a circle around the ecliptic pole of the solar system. Polaris will be nearest the pole (about ½ degree) at the end of this century.

The last time there was a star at all near the north pole was prior to the construction of the Great Pyramid in 2540 BCE. That pole star was Thuban (Alpha Draconis), and it was just one-fifteenth of a degree from the pole in 2800 BCE. The Great Pyramid has a narrow air shaft that pointed to Thuban, at which time it was already departing the pole and nearly one-third of a degree from it. Therefore, the megalithic people at Carnac, as well as almost all cultures throughout time, did not have the convenience of a pole star in approximately locating the north pole.

From 5000 to 4000 BCE, the time of megalithic building at Carnac, the north pole was a dark region surrounded by many bright stars. The inability to locate the north pole using a pole star challenged the people of the megalithic to develop a more sophisticated and accurate method. In any case, true north and latitude needed to be located more accurately than by using a pole star, which can only ever approximate the position of the north pole. Also, through the circumpolar observatory, sidereal time and even longitude between sites could be measured once the movement of the circumpolar stars could be exploited. True north, based upon these stars around the pole, can give the cardinal directions to an observatory.* Toward the end of the megalithic study of astronomical time, methods such as day-inch counting would have proved insufficient to resolve the small fractions of a day required to quantify, for example, the full behavior of the moon. This need for greater accuracy led to the full development circumpolar observatories.

Circumpolar astronomy can make measurements leading to an accurate estimate of the size of the Earth. This may explain why the end of astronomical *research* at Carnac turned into the physical *search* to find the size and shape of the Earth, in England and then Egypt. In Egypt there are clear results of this geodetic work in the dimensions of the Great Pyramid (see later section, "Early Perspectives on a Created World"). In southern Britain the geodetic research led to another monument built at the same time as the Great Pyramid and presenting more results from the geodetic work within its dimensions: the Sarsen Circle of Stonehenge.

---

*The equinoctial sunrise in the east and sunset in the west can give a mean azimuth (horizon angle) to obtain south and north but only if the horizon is dead flat to each of these alignments.

Before investigating the work in Britain in chapter 6, the operation of circumpolar observatories needs to be explored as both the pinnacle of astronomical enterprise and the essential means for the megalithic astronomers to later realize their model of the Earth. Only through the lens of circumpolar astronomy has the greatest observatory of this type, the western cromlech at Le Menec, become intelligible.

HOW CIRCUMPOLAR OBSERVATORIES WORK

Looking north, the megalithic circumpolar observatory could track a circumpolar star, circling the pole in a counterclockwise direction as the Earth rotates forever eastward. Because a given marker star describes a circle of fixed distance from the pole, the horizon angle (or azimuth) of the star, when brought down vertically to form a horizon alignment, will achieve a maximum elongation on either side of the north pole. When these elongations are measured as two alignments on the horizon, to east and west, the midpoint between these two alignments will give a definite bearing for true north.

Once the maximum elongations of a marker star have been established relative to an erect pole used as *foresight*, their sightlines can run past the pole and continue south in the form of an A shape and, as the A expands south, points at an equal distance from the pole (at the vertex of the A) can be chosen so as to form the diameter of a circle (see figure 4.5). This circle can then represent the circumpolar orbit of the chosen marker star, in azimuth, throughout its daily motion.

As the marker star moves between its two extremes, an alignment to the star can be updated as a point on the backsight line, which marks the circle's diameter between the eastern and western elongation points. Because east and west become reversed as you cross the foresight, to restore east and west when creating the circumpolar orbit circle, the star's location on the backsight line must be transferred across the center point to the opposite side. Then the location of the star in its circumpolar orbit can be located perpendicular to the east-west diameter backsight line, on a point on the circumference circle directly north or south of the backsight line (according to whether the star appears above or below the altitude of the north pole).

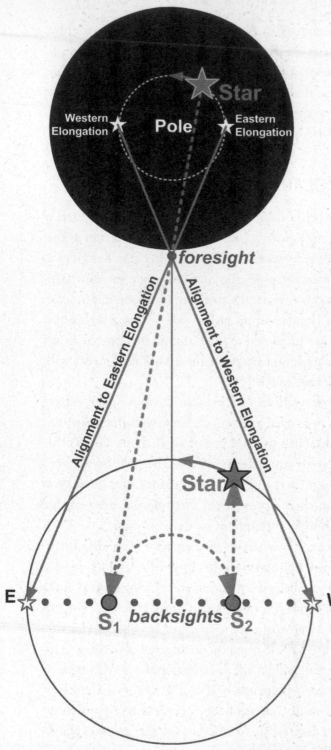

Figure 4.5. How a simple circumpolar observatory works, based upon foresight to azimuth alignments of a chosen marker star. The backsight line is established between the eastern and western alignment points, marking the diameter of the circle that represents the circumpolar orbit of the marker star. Because east and west become reversed as you cross the foresight, to restore east and west the star's location, S_1, must be transferred symmetrically around the center of the circle to become S_2. A star marker can now be placed on the circumference of the circumpolar orbit circle at the same angle as the star relative to the circumpolar sky.

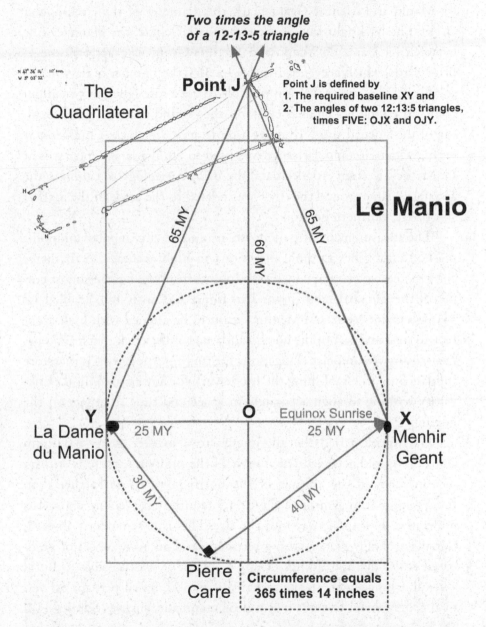

Figure 4.6. A possible early circumpolar observatory at Le Manio. The circumpolar circle's diameter is marked by Menhir Geant and La Dame du Manio, which are exactly 50 megalithic yards apart. This circumference is evenly divisible by 365, the number of days in a solar year, units of 14 inches.

An example of such an observatory is found at Le Manio, where the northern point of the Quadrilateral marks the foresight and La Dame du Manio and Menhir Geant mark the diameter of the circumpolar circle. The backsight line of the circumpolar circle, the diameter line between La Dame du Manio and Menhir Geant, is exactly 50 megalithic yards, which is evenly divisible by 365, the number of days in the solar year. This shows that it is possible to use a foresight and two alignments to establish a circular clock on the ground that will give a reading of the sidereal time. To make a more accurate clock, a marker star with a wider circumpolar range was used to build a larger apparatus at Le Menec's western cromlech and, by using metrology to calibrate the apparatus, time around the circle was related to the angle of the marker star on the northern horizon.*

The measurement of true north using the circumpolar midpoint method establishes cardinal directions to high accuracy. A diameter constructed between equally distant points from the northernmost corner of the Quadrilateral (point J in figure 4.6 on p. 89), enabled Le Manio's Dame stone and Menhir Geant to be aligned with high accuracy to west and east of the north-south axis of the circle. Such an apparatus, using circumpolar elongations for finding true north, is therefore fundamental to establishing the framework for any astronomical observatory wishing to align accurately in space and time to events on the horizon.

Having identified the alignments to a marker star's maximum elongations and achieved true north as the midpoint, there is still no measurement of the altitude of the north pole. To go further than the gnomon arrangement of figure 4.5 requires an accurate apparatus point in a measurable way into the sky. This is how modern observational astronomy works, with a telescope and an accurate set of graduated scales giving readings of angle. While it was possible to build vertical arrangements in the megalithic period, based perhaps on vertical geographical features (natural or manmade, such as Silbury Hill

---

*This is remarkably close to our modern concept of the hour angle, which forms the basis in the sky of right ascension and which defines the point on the celestial equator that rises with any celestial object.

Figure 4.7. La Dame du Manio (above), the point of maximum elongation to the east for marker stars Alioth and Pherkad and marking the westernmost point of an observatory circle, diameter 50 megalithic yards. Menhir Geant (left) marks the eastern point, here shown at the 2010 spring equinox sunrise.

near Avebury), it was quite difficult. Perhaps it was done where larger civilizations enabled pyramidal building. As we will see in part 2, the circumpolar observatory had ways to define latitude through relative longitude.

Finally, it should be noted that when one studies the change of angle of circumpolar stars over time, one soon notices that some stars move very rapidly (within a few decades) near the north pole, either toward it or away from it. This fact must have led to the realization that it is the pole as a single entity that is moving relative to all the stars, which only appear to move. The different movements of the stars relative to the pole and the equinoctial and solstitial points are the result of the location of those stars relative to the pole in distance and angle. This would have been a tremendous conceptual advance, a move from seeing many phenomena of movement and realizing instead a single cause, the motion of the Earth's axis. We recognize such a profound collapse into simplicity as a new understanding and a change of paradigm, after which the world is a different place. This profound progress by the megalithic astronomers explains how it was *possible* for the knowledge of precessional time—the time taken for the stars to return to their same relative positions in the sky—to have entered the mythic record.

CAPTURING SIDEREAL TIME

We can now complete our treatment of Carnac's astronomical monuments by returning to Le Menec where the challenge was to measure time accurately in units less than a single day. This is done today at every astronomical observatory using a clock that keeps pace with the stars rather than the sun. The 24 hours of a sidereal clock, roughly four minutes short of a normal day, are actually tracking the rotation of the Earth since Earth rotation is what makes all the stars move. Even the sun during the day moves through the sky because the Earth moves. Therefore, in all sidereal astronomy, the Earth is actually the prime mover.

The geometry of a circumpolar observatory can reveal not only which particular circumpolar star was used to build the observatory but also the relatively short period of time in which the observatory was designed. Each bright circumpolar star is recognizable by its unique elongation on the horizon in azimuth and its correspondingly unique and representative circumpolar orbital radius in azimuth. Around 4800 BCE, Alkaid, the star at the end of the Big Dipper, had an elongation equal to the diagonal of a triple-square, and this date corresponds to the carbon dating of parts of the Locmariaquer complex, a site designed upon the triple-square. The entrance to the Table des Marchands dolmen appears to hold, on its left-hand jamb, a carving of the Big Dipper, shown in figure 4.8b on page 95. The graph in figure 4.8a on page 94 shows the movement in maximum elongation of Alkaid over the fifth and fourth millennia BCE and one can see that Alkaid only achieved the triple-square azimuth during a small window of about one hundred years, from 4800 to 4700 BCE.

As discussed above, the 50-megalithic-yard separation of La Dame and Menhir Geant at Le Manio was perfect to develop an observatory circle with 365 units around a circumference representing the "orbit" of the circumpolar star. But again, the geometry used was fixed, being that of a 12-13-5 triangle and useful only for circumpolar stars with a maximum elongation of 22.6 degrees, to east and west of north. Two stars were suitable for this geometry, Alioth (epsilon Ursa Major) around 4236 BCE and Pherkad (gamma Ursa Major) around 4340 BCE, so that Pherkad and then Alioth could have provided continuity at the observatory from 4400 to 4200 BCE.

By the time Le Menec was built, a new type of geometry had been developed, useable with almost any circumpolar star. The 40-degree elongation of Dubhe (alpha Ursa Major) perhaps was chosen there to give a broader angular movement upon the northern horizon. The three observatories we have discussed, Locmariaquer, Le Manio, and Le Menec, appear to have formed a continuity over all of the fifth millennium BCE, see figure 4.9 on page 96. Over the course of this epoch we can see vast improvements in the design of the circumpolar observatory.

Sculptures of axes have been thought to represent Ursa Minor, but it seems more probable that any circumpolar stars chosen as a marker star could be given this motif, especially the brighter stars of Ursa Major. The axe symbol was also the axe plough or *ard* of the Neolithic and has become one of our enduring names for Ursa Major, "the Plough," which refers to Ursa Major's bright stars (which now look like a more modern plough due to movement of the pole). It was probably Alkaid that ploughed the circumpolar heavens once per sidereal day on the ceiling of Table des Marchands (see figure 4.1 on p. 80).

Knowledge of precession is not thought to have existed in prehistory, but megalithic astronomers, who looked at the circumpolar stars and tracked a specific star's maximum elongation in azimuth, would

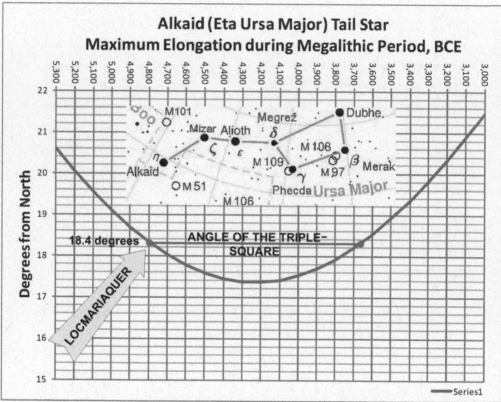

a

Figure 4.8. (a) (above) The variation in elongation of circumpolar star Alkaid gives a date at Locmariaquer of 4800 BCE, when the triple-square to north would have operated with Alkaid as circumpolar marker star. (b) (opposite) The "question mark" glyph on the left-hand jamb in Table des Marchands (seen on the left) can be interpreted as Ursa Major, which is where the sun at midsummer rises when this constellation stands to the east of the north pole (see right). Also shown on the right is the geometry of a three-by-four square, which has a diagonal of five, at the summer solstice 3-4-5 angle. Ursa Major did "stand up" in the northern sky in this way, to the right of the pole at this epoch, when that part of the ecliptic was rising on the eastern horizon.

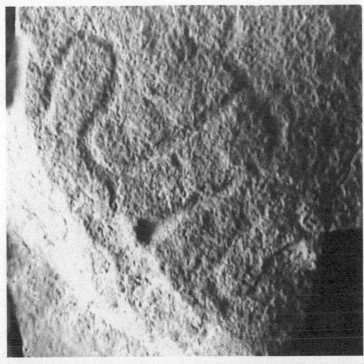

b

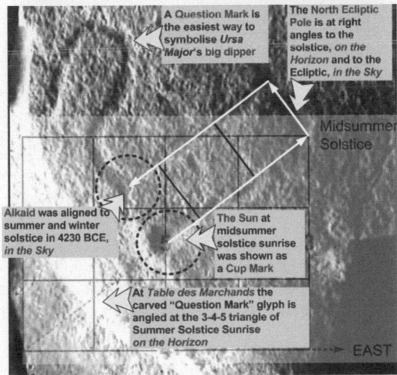

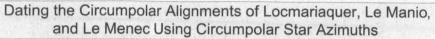

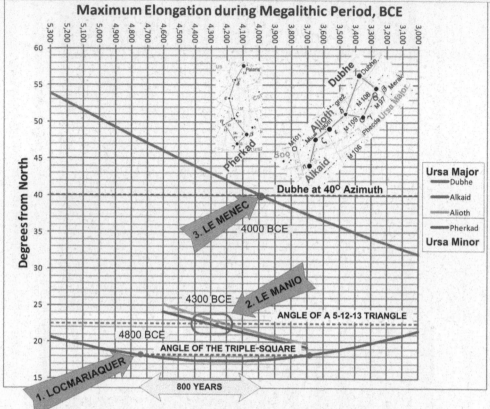

Figure 4.9. How a circumpolar observatory allows an extraordinary way for dating its useful life. Because the pole moves so fast relatively to nearby circumpolar stars, we can accurately date megalithic circumpolar observatories. The three observatories we have discussed would have a chronology of Locmariaquer around 4800 BCE, Le Manio around 4300 BCE, and Le Menec around 4000 BCE.

have surely discovered precession. Because circumpolar stars are so close to the pole, they are therefore rapidly changing their declinations and their maximum elongations in azimuth, thus exaggerating the phenomenon of precessional change.

Le Menec circumpolar observatory has some new alignments to consider when one includes some of the circumpolar stars, in particular the "western alignments," the parallel rows of stones set to the east of the circumpolar orbit circle. The western starting stones of rows 6 to 9 of the western alignments form a line that points north-north-east.

The cromlech design is based upon a circle, which was subsequently enlarged into an egg shape, known in British monuments as Type 1. But the observatory part was based upon the original "forming circle" used to develop the egg shape (see figure 4.10). The east-west diameter of the forming circle, defined by Dubhe's maximum elongations, can also be seen to define the base of a rectangle, with the short side defined on the

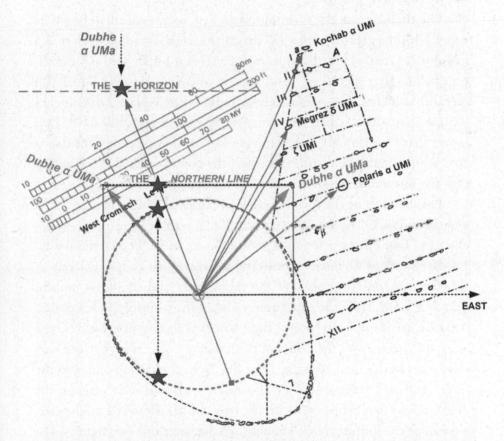

Figure 4.10. The key operation within Le Menec's western cromlech, enabling Dubhe to be brought down to Earth as an "hour angle" and thus track sidereal time. Based on the western starting stones of Le Menec's parallel rows of stone, the western alignments, it appears other stars were also considered. The leading stone of row 6 aligned to Dubhe as marker star and with the leading stones of rows 7, 8, and 9 formed a line that points north. Note that row 9 starts with a tryptic of three stones. Diagram drawn according to Thom's survey.

eastern edge by the leading stones of rows 6 through 9. The starting stone of row 6, the northeastern corner of the rectangle, is aligned to the maximum elongation of Dubhe. The diameter length is therefore repeated to the north and, using sightlines through a central pole, the azimuth of the star Dubhe could be located upon this new "Northern Line." Once located, a bearing directly south will touch the circumference of the forming circle in two places and one of these will be the visual position of Dubhe as seen in the circumpolar sky, but is now measurable upon the Earth.

The diameter of the forming circle had been carefully chosen as being 3,400 megalithic inches (17 megalithic rods in radius, each of 2.5 megalithic yards). There is evidence for such a 40-inch yard (by definition within later metrology), and a "rod" of 2.5 would then equal 100 megalithic inches. Seventeen rods in radius then equals 3,400 inches in diameter, which is half of a lunar nodal period of 6,800 days. This meant that a continuous inch count could start on one end of the diameter, at maximum standstill, and when the count reached the opposite end, the minimum standstill would be occurring.

The diameter of the forming circle in regular inches was then 2,773 and the radius in regular inches was 1,386½, significant as the number of days in four eclipse years, the Octon eclipse period. The Octon is only 1.4 days short of 47 lunar months, which is why an eclipse is likely at this first near meeting place of the eclipse year and the lunar month. Traveling north from the easternmost end of the forming circle's radius, one soon arrives at the tryptic of three stones at the start of row 9. These are 1,461 inches from the center of the forming circle, the length of four solar years in day-inch counting. The line from the circle's center to the tryptic forms the diagonal of a triple-square with four solar years on the diagonal and four eclipse years on the base, which is formed by the eastern half of the forming circle's established diameter line (see figure 4.11).

The outer circumference of the observatory was based upon a radius equal to the circumference of the lunar simulator at Le Manio, 1,394 inches, because this gave a perfect division of the circumference into 365 units of 24 inches each. Dividing the circle by 365 provided units of angular motion equal to the sun's average motion each day on the ecliptic. The units of 24 inches strangely echo the number of hours cur-

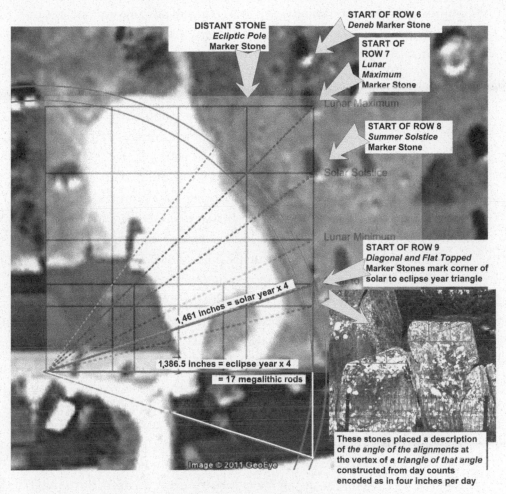

Figure 4.II. The now intelligible organization of the stones in the northwest sector of the western cromlech. The triple-square is terminated by a tryptic of stones clearly labeling the angle chosen for the western alignments as that of the diagonal of a triple-square. Main image adapted from Google Earth and GeoEye, bottom right inset image from *Carnac: The Alignments* by Howard Crowhurst.

rently given to a sidereal (or solar) day. The sidereal time of day would be the location of Dubhe on the forming circle after it had been transposed from the sky onto the Northern Line and then onto the observatory circle.

At any given position for Dubhe on the circle, exact parts of the sky were rising in the east and setting in the west and therefore very specific points on the ecliptic were rising or setting. Knowing where the sun would be rising, if it was at the same location as the moon marker on the lunar simulator, at last gave the astronomers the ability to measure the difference in location and timing of the moonrise, relative to where the sun would have risen at the moon's ecliptic longitude. This then isolated the amount by which the moon deviated from the ecliptic in its orbit, and *this* became the subject for the western alignments, the "swan song" of the astronomical works there.

The variation of stones in the same rows of the western alignments indicated the systematic drift of ecliptic latitude for moonrises, aggregated into portions of the ecliptic. Using a mixture of Google Earth imagery and a Thom plan of the alignments (with a scale of one mm to one meter), this wandering of the rows can be analyzed as shown in figure 4.12.

The stone rows appear to have measured the variation of the moon's ecliptic latitude during the half cycle belonging to the major standstill. Each western alignment row belonged to a region of the ecliptic as it rose, the ecliptic latitude in that region slowly changing. The results do not appear as very significant, but at least there is now an explanation for the variations in the rows.

The knowledge that was discovered due to the Le Menec observatory is awe inspiring when the perimeter of the egg shape is taken into account. It is close to 10,000 inches, the number of units of sidereal time the moon takes to orbit the Earth. The egg was enlarged in order to quantify the orbit of the moon as follows: every 82 days (three lunar orbits) the moon appears over the same part of the ecliptic. Dividing the ecliptic into sidereal days we arrive at 366 units of time per solar day.* 82 days times 366 divided by the three lunar orbits gives the moon's sidereal orbit as 122 times 82 day-inches. Instead of dividing 82 by three as we might today to find the moon's orbit, the pre-arithmetic of metrology enabled the solar day (of 366 units) to be divided into three lengths of 122. If a rope 122 inches long is then used 82 times (a whole number), to lay out a longer length, a length of 10,004 inches results. If

---

*These units are each the time required for an observer on the surface of the Earth to catch up with a sun that has moved within the last 24 hours, on the ecliptic, a time difference of just less than four minutes.

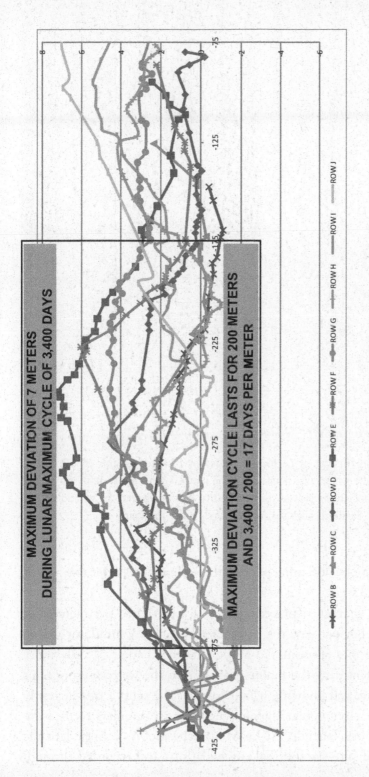

MAXIMUM DEVIATION OF 7 METERS DURING LUNAR MAXIMUM CYCLE OF 3,400 DAYS

MAXIMUM DEVIATION CYCLE LASTS FOR 200 METERS AND 3,400 / 200 = 17 DAYS PER METER

—×—ROW B —+—ROW C —▲—ROW D —■—ROW E —✳—ROW F —●—ROW G —+—ROW H —ROW I —ROW J

Figure 4.12. Analysis of the wandering of stones along the rows of the western alignments. Data from Thom survey, captured by David Blake.

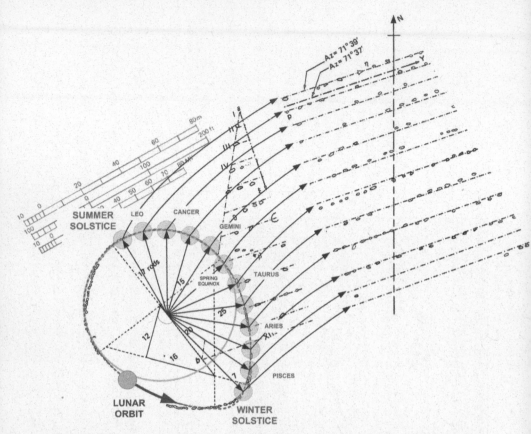

Figure 4.13. The perimeter at Le Menec divides by 82 into sections of 122 inches. This meant it could function as a simulator of the moon, connected to both the ecliptic (through the egg) and to the part of the ecliptic that was rising with the rotation of the Earth (through the forming circle's 365 divisions).

10,004 is divided by 366 units per day then the moon's orbit emerges as 82/3 or 27⅓ days.

If a moon marker is placed upon the Le Menec perimeter and moved 122 inches per day, the perimeter becomes a simulator of the moon, like the one encountered at Le Manio (chapter 3) only more sophisticated. Knowing the moon's position on the western cromlech's model of ecliptic and knowing which parts of the ecliptic are currently rising from the circumpolar stars enabled the astronomers to measure the moon's ecliptic latitude. At Le Menec the moon's ecliptic latitude appears divided into six zodiacal constellations and recorded through

11 to 12 stone rows (the Alignments), starting near the egg-shaped perimeter and forming the western alignment rows as a series of measurements, within the moon's nodal period of 18.6 years. As successive moonrises marched from south to north during half of each lunar orbit (see figure 4.13), the moon's ecliptic latitude above or below the sun's path could be recorded by stones as their deviation from the general bearing of their row, relative to where the sun would rise at the same ecliptic longitude (see figure 3.30 on p. 46).*

EARLY PERSPECTIVES ON A CREATED WORLD

The attribution of will to celestial objects, such as in astrology, was a type of religious outlook that dominated early civilizations, but its nemesis appears to have been the rise of Christianity. When the Church of Rome was intellectually marginalized through the rise of modern scientific thinking, the latter eliminated any vestiges of astrological speculation as part of the former spiritism that had obscured the physical causation behind natural forces. However the "sister" discipline of astronomy, which named, cataloged, analyzed, and established the physics of the universe, was retained. Two intellectual veils have therefore been cast over the ancient notion that the cosmos might be participating in some way with our lives on Earth. Neither Christians nor scientists think that the organization of the world serves a higher purpose through its highly specific architecture. This places Egyptian theological ideas of the sacred geometry of the Earth and heavens, like the notion of advanced megalithic science, in a "no man's land" of pure fantasy.

If ancient cosmogenesis was a unification of religion and science within the consciousness of prehistoric humans, then some portion of an unrecorded megalithic theology might have leaked into ancient religious texts. This would be especially relevant if, as I propose, the Egyptians and megalithic Europeans cooperated to survey their respective

---

*No longer "stones marching in stiff-shouldered lines, marching nowhere, doing nothing," as Magnus Magnussun put it in *Cracking the Stone Age Code,* a BBC documentary from the 1970s about Alexander Thom's work, including at Carnac (available online).

sections of the Earth's meridian. Some Egyptian religious thinking has survived that might express some of the religious thinking arising alongside the great monuments of Carnac. (We have already shown, in chapter two, how Egyptian geometrical calculation could have evolved from the system of multiple-squares, which also appears at Carnac.)

The primary justification for a religious transmission to Old Kingdom Egypt would be its association with the metrological transmission that must have occurred prior to the measurement of the different degrees of latitude along the Nile valley. The window of opportunity for any such transmission would begin no earlier than 3400 BCE (if the people of megalithic Europe were to be involved with those of archaic Egypt) and end no later than 2800 BCE since the Great Pyramid, dated at 2540 BCE, contains dimensions that correspond to a geodetic survey of Egypt. This dimensional information is fairly primary to the design of the Great Pyramid: its height multiplied by its four different side lengths generates a rectangle whose area in square feet equals the number of linear feet found in four of Egypt's latitudinal degrees.

The northern side of the pyramid, 755.765 feet, multiplies with its full height of 481.09 feet to give a length encoded as an area of 363,427.5 square feet in what is called a rectangular number. According to F. R. Helmert's 1907 work on the best geodetic ellipsoid for Egypt, this length is that of the degree south of 26 degrees, the latitude of Karnac (it being located at $2/7$ of the meridian or $25\frac{5}{7}$ degrees north).[1] In similar fashion, the other sides give the exact latitude lengths for the three consecutive degrees surrounding the pyramid's latitude of 30 degrees, using the common factor of its capped height. This rectangular encoding of four latitudes using the pyramid's side lengths only generates latitudinal degree lengths when the measurements are in simple English feet throughout, so this unit of length must have been used by the builders when designing the pyramid in order to obtain this result. The English foot is a unit first detectable in Carnac's monuments between 5000 and 4000 BCE and in megalithic Britain from 3500 to 3000 BCE. The most famous Egyptian measure, the Royal Cubit of 12/7 feet, was engraved within Gavrinis, east of Carnac by 3500 BCE.

The first Pharaoh, Narmer (ca. thirty-first century BCE), was the

RECTANGULAR AREA = H$_p$ *times* S$_{north}$
= Length of the Latitude of Karnac

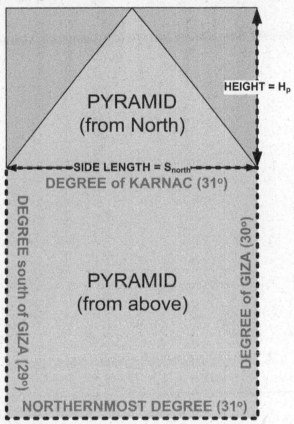

Figure 4.14. Example of how the base lengths of the Great Pyramid were built to give the lengths of different degrees of latitude within Egypt when each is multiplied with the pyramid's height. Illustration based on John Neal, *All Done with Mirrors*, pp. 134–37.

first to unify the different kingdoms of Upper and Lower Egypt—a story told in the iconography of the Narmer Palette. Narmer's reign laid the foundation for later dynasties, and the zenith of the Pyramid Age was reached within five centuries, by the fourth dynasty, in the building of the Great Pyramid and the Giza complex (2540 BCE).

The height of the Great Pyramid, already used to display latitudes relative to slightly different side lengths, was reused to show the ratio between the polar radius of the actual Earth and the polar radius of the

Figure 4.15. Both sides of the Narmer Palette, which contains symbols of the First Dynasty unification of Upper and Lower Egypt. Photos from www.francescoraffaele.com.

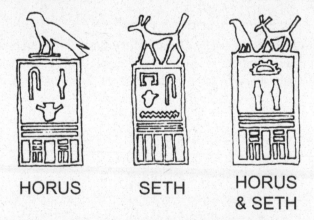

HORUS   SETH   HORUS
                         & SETH

Figure 4.16. The successive "labels" of three pharaohs during the turbulent Second Dynasty, showing how the hawk of Horus was replaced with the doglike Seth only for both to become integrated. From *The Sphinx and the Megaliths* by John Ivimy (London: Turnstone, 1974).

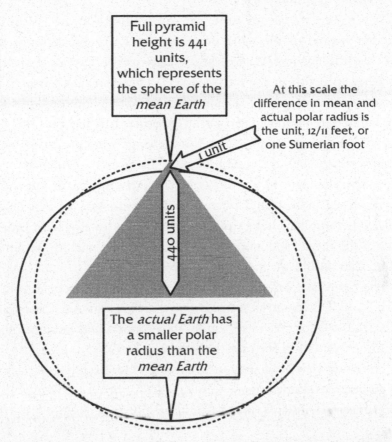

Full pyramid height is 441 units, which represents the sphere of the *mean Earth*

At this scale the difference in mean and actual polar radius is the unit, 12/11 feet, or one Sumerian foot

1 unit

440 units

The *actual Earth* has a smaller polar radius than the *mean Earth*

Figure 4.17. The vertical version of the megalithic model of the Earth, showing the difference between the actual polar radius and the mean Earth radius as the difference between the capped and uncapped height of the Great Pyramid.

mean Earth.* The ratio between the height of the uncapped pyramid relative to its capped height is 440 to 441 and the difference in height is 1.09 or 12/11 feet, historically called the Sumerian foot. By leaving the actual pyramid uncapped, thus representing the actual Earth, the unit of difference between the two radii was made visible as being 12/11 feet. The uncapped height of the pyramid is then 440 times 12/11 feet or 480 feet.

---

*As discussed earlier, the Earth's rotation causes the equatorial diameter of the Earth to be larger than the polar diameter. The "mean Earth" represents the dimensions Earth would have if it were a perfect sphere.

The Sarsen Circle of Stonehenge (2600–2400 BCE) presents a related version of the same model of the Earth by demonstrating how the different metrological units relate to the Earth's dimensions (described in depth in the next chapter). Therefore, the Sarsen Circle contained one statement of the metrological system and the Great Pyramid, built contemporaneously with it, another. It is most unlikely that these two monuments at Giza and Stonehenge were *not* based upon a shared technology and shared endeavor. There is no evidence for any original development of metrology by the early Egyptians but there is such evidence in Brittany. Therefore, a transmission of metrological knowledge must have taken place from Europe to Egypt in order to produce the megalithic model of the Earth found within monuments and implicit within the metrological system.

The Egyptians certainly knew Hyperboreans as "people from the north," and a high-status statue from the Pyramid Age, "the Seated Scribe," has eyes of green using glass with copper. Did he signal northern influences within the priestly classes?

The world view of the Egyptians around 1200 BCE placed their limited geographical knowledge within a planetary *circle,* or "island" Earth, in which certain lines of latitude could be shown (see figure 4.19). This

Figure 4.18. The so-called Seated Scribe on display in the Louvre, Paris, originally from Saqqarah, Fourth or Fifth Dynasty, 2600–2350 BCE. Made of painted limestone, the statue has eyes inlaid with alabaster, quartz cornea, and rock crystal irises set in copper to render them green, perhaps signifying northern influence in the priestly classes of Egypt.

map caused Herodotus to write in the mid-fifth century BCE in his *Histories*: "And I laugh when I see that, though many before this have drawn maps of the Earth, yet no one has set the matter forth in an intelligent way; seeing that they draw Oceanus flowing around the Earth, which is circular exactly as if drawn with compasses, and they make Asia equal in size to Europe . . ." Herodotus did not realize how ancient maps had

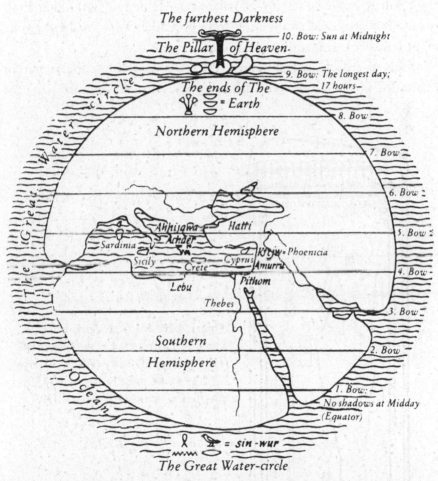

The furthest Darkness

The Pillar of *Heaven.*

10. Bow: Sun at Midnight

The ends of The = *Earth*

9. Bow: The longest day; 17 hours—

8. Bow

Northern Hemisphere

7. Bow

6. Bow

Ahhijawa — *Hatti*

5. Bow

Sardinia *Achaer*

Sicily *Crete* *Cyprus* *Kftiu* - *Phoenicia*

Amurru

4. Bow -

Lebu *Pithom*

Thebes

3. Bow

The Great Water circle

The 'Ocean

Southern Hemisphere

2. Bow

1. Bow: No shadows at Midday (Equator)

= *sin -wur*

The Great Water-circle

Figure 4.19. The Egyptian world picture circa 1200 BCE. The "circle of the Earth" is surrounded by the "great water circle" (sin wur, Greek Okeanos). The circle of the Earth is divided into nine "bows," hence the whole world can be described as "all the nine bows." The ninth bow lies "at the ends of the Earth in the furthest north." The Greeks called the "heaven pillar" stele boreis, which means "north pillar," and the heaven bearer, Atlas. From *Atlantis of the North* by Jurgen Spanuth (London: Sidgwick & Jackson, 1979), p. 29.

been a convenient combination of geography and geodetic cosmography.

The map on p. 109 shows "pillars of heaven" at the far north, holding up the north pole. The pillars were immortalized by the Greeks as "Atlas," who holds up "the world" of the celestial Earth. For the Egyptians the Earth was *Tatanen* (the Primordial Hill), who had been exalted by *Ptah,* the Source of all that exists. This is found in the Memphite theology, where *Ptah-Tatenan* is the pillar bearing the whole of the rotating heavens. Jeremy Naydler says, "It is significant that Ptah is usually represented as a mummiform god, like Osiris, with whom he eventually became identified."[2]

Ptah, as represented in figure 4.20, is one-legged like the Sumerian fish-tailed God Oannes and Indian prince Dhruva, immortalized for

Figure 4.20. The mummiform god Ptah in a shrine as depicted in the Papyrus of Ani, Eighteenth Dynasty (1250 BCE). From *Temple of the Cosmos* by Jeremy Naydler (Rochester, Vt.: Inner Traditions, 1996).

standing on one leg, as the polar axis.[3] Sumerian Utnapishtim, who built the first Ark, was immortalized to live "where the two rivers meet"—that is, at the equinoctal point where the celestial equator is crossed by the sun's ecliptic path. This metaphor works well with Mesopotamian geography where Eden would have had four rivers rather than two.

Naylor continues, "The Memphite theology thus relates to that aspect of the divine that is most involved with matter, the aspect that has sacrificed a purely spiritual mode of being in order to be crystallized in materiality."[4] Thus, Ptah-Tatanen could be seen as a composite of the Aten's power to create the world.

> *The mighty Great One is Ptah,*
> *who transmitted life to all gods,*
> *as well as to their kas*
> *through this heart, by which Horus became Ptah,*
> *and through this tongue, by which Thoth became Ptah.*[5]

Naydler says, "The Divine Company, or Ennead, is in this way distinguished from the Heliopolitan Ennead. For the Heliopolitan Ennead arose as a result of Atum's act of procreation, whereas the Memphite Divine Company is identified with the [ongoing] act of procreation."[6]

> *. . . he [Ptah] is [as heart] in every body, [as*
> *tongue] in every mouth, of all gods, people, beasts,*
> *crawling creatures and whatever else lives . . .*[7]

The act of creation is continuing through all beings, seen as having a heart and a tongue, as Ptah. Horus was naturally associated with the sun as Horus-Ra, the heart being the seat of intellect. Thoth was associated with measure and the moon. I believe Ptah-Tatanen represents the Earth and its pole, especially in regard to the organization of time on Earth.

Continuing with Naydler, "The Memphite cosmology presents us with the fulfilment of the divine creative process, the final embodiment of the divine substance in material form. Viewed in this light, the cosmogonies of Heliopolis, Hermopolis, and Memphis do not appear as rivals so much as complementary aspects of a greater cosmogonic scheme . . ."[8]

The two active aspects of creation can be related to the sun and equator (Horus) and the moon and meridian (Thoth). The paths of the sun and the moon both deviate from the axis of the Earth. The sun is skewed due to the axial tilt of the Earth's rotation, giving the Earth its seasonal and latitudinal variations in climate and habitat so crucial to life. The moon is skewed relative to the sun's path, a double deviance, but without which the creation and ongoing evolution of the moon's relationship to the Earth, and life itself, would not have existed because the moon stabilizes the Earth's tilt and its precessional cycle while achieving many other things for life on Earth.[9] We have seen the moon giving the Stone Age its earliest lessons in astronomical time, from bone tallies to Carnac's monuments and the three-square and four-square diagonal relationships between different years, only meaningful because of the moon's periodicities relative to the sun.

The moon has the Indo-European root of *me,* which is primary to "measure." Thoth was identified as being a measuring principle within creation, like Set (or Saturn) who cut up an *unrealized* generative power, the cosmic ocean Okeanus, which surrounds the Primordial Hill of Earth. The point here is that astronomical understandings probably underlay the later theological developments of Egyptian religion and are most likely to have done so through knowledge transmitted from the megalithic peoples of Europe, especially as already established, from the megalithic builders at the sites of Carnac.

PART TWO

The Journey Back to Earth

The Kerloas Menhir, near Plouarzel. With a height of 9.5 meters this menhir is the tallest standing menhir in Brittany A few centuries ago the top was destroyed in a thunderstorm. Originally it must have been over 10 meters high. Many such menhirs could have been part of observatories that established the latitude and longitude relative to other similar observatories, with the rest probably functioning as milestones to mark the distances between "places" (i.e., established observing sites).

5

LOOKING SOUTH TO
MEASURE THE EARTH

WHILE CARNAC IS world famous as a megalithic district, much of Brittany demonstrates megalithic activities over the one degree of latitude from Carnac to the Breton northern coast. The altitude of the polar axis can be seen to change depending on the latitude of the circumpolar observatory. The megalithic scientists in Brittany would naturally have investigated the relationship of their different latitudes, and the evidence shows that they did this by measuring ground lengths between them. This would explain the move from Brittany to Britain (perhaps via Ireland) later in the fourth millennium when they produced a simple geodetic model of the Earth. This model underlies ancient metrology and enabled monuments like the Sarsen Circle and Great Pyramid to contain the size and shape of the Earth within their dimensions.

Having a developed metrology and geometry, the megalithic people in Brittany began to see that the Earth was not a perfect sphere. Lunar simulators and circumpolar observatories had already revealed that the Earth was some sort of spinning globe, but it would have been unclear why distances grew longer when moving north between sites the same difference in latitude apart. This small lengthening of equal differences in latitude as one moves north would have been recognized when observatories all over Brittany became linked via a geometrical grid (see figures 5.1 and 5.2). The megalithic.co.uk database shows hundreds of singular standing stones or menhirs dotting the landscape, and some of these share the same latitude. The distance between these menhirs could have been located according to their distance on the ground, their angular bearing, and their latitude so as to create a picture of changes in the sky relative to locations on the landscape.

Indeed, similar networks of stones were often erected in the ancient world, as in Greece where they functioned as simple wayside markers indicating the distance from that point to everywhere else. Another example is the Temple of Amun at the latitude of Thebes in Egypt. When it was capital of a unified Egypt, it was the geodetic center from which all distances were measured. The ability to combine geometry, latitude, and distance, even over reasonably small portions of the globe, revealed that the Earth was not a perfect sphere. By looking at stars transiting to the south, it was possible to use a circumpolar observatory

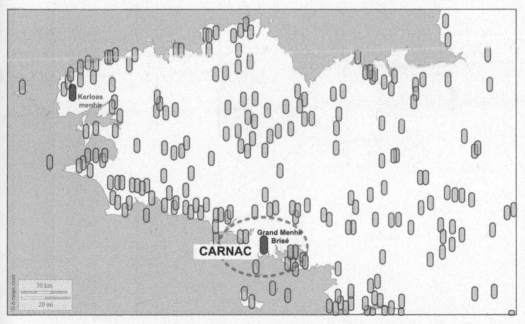

Figure 5.1. The number of surviving menhirs in Brittany are legion, and this picture shows just some of those that have survived. These could have provided information about the Earth's shape and launched a new area of geodetic studies, requiring a journey to southern Britain and possibly Egypt. Data courtesy of megalithic.co.uk, outline from d-maps.com.

in a new way, to measure both differential longitude between two sites and their exact geocentric latitude.

The natural way to divide up a circumpolar observatory circle is to divide its circumference into 365 parts, to conform to the solar motion per day. At Le Manio the radius of the circumpolar observatory is 25 megalithic yards, giving a 365-unit circumference where each unit is 14 inches long, accurate to one part in 8,000. At Le Menec the radius is 1,394 inches (17 times 82), giving a circumference where each of the 365 units is equal to 24 inches, to the same accuracy. This division by 365 has great bearing upon the establlished size of the English foot— the equator is 365.222 times 360,000 feet long.

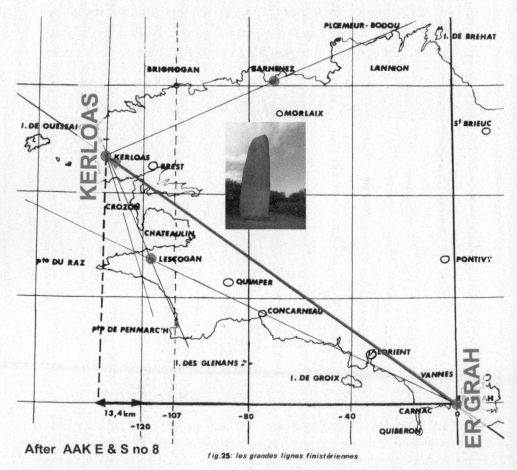

After AAK E & S no 8

fig.25: les grandes lignes finistériennes

Figure 5.2. In the 1980s the AAK discovered geodetic relationships between sites in Brittany.

Two circumpolar observatories on the same latitude can roughly determine their difference in longitude through the difference in angle of their marker stars when a lunar eclipse begins. In the same way, the stars of the southern sky seen from two different longitudes will be angled differently because of the difference in longitude between the sites. Effectively, they are in slightly different "time zones," something ignorable today until one goes over a time zone boundary. If both sites used the Le Menec design a couple of degrees apart in longitude, the angular difference in the same marker star at the onset of an lunar eclipse would be about 48 inches apart (1/180 of the parallel), a fraction of their 365-unit circumference. Their angular measurements to the marker star might only be accurate to 1/20 of a degree but many readings would be able to give an average, the onset and disappearance of the shadow for each eclipse giving two readings.

But how could the scientists of the megalithic already have an inch and then a foot of twelve inches with this relationship to the Earth? The answer probably lies in yet another coincidence aiding the people of the megalithic. If the thumb is pressed against a ruler, the joint is an inch across. This is why the French call the inch *le pouce*, or "thumb." It is possible that the inch was standardized in Brittany from a measurement of one or many thumbs and then used in the astronomical work at Carnac. This standardized measure would have led to a foot of twelve inches, the English foot, due to a lunar year being seen as twelve times the number of day-inches in a lunar month. Therefore, it is possible that the inch existed because of the regular size of the local human thumb, only coincidentally supplying a nice number of feet for the length of the equator. (After all, your thumb could divide the equator into 365¼ parts, each approximately 4,320,000 thumb widths long.)

At latitude 48.189 north, the latitude of megalithic sites in Britain, the Earth's radius is two-thirds that of the equator and therefore the Earth's circumference along that parallel is two-thirds that of the equator, namely, 365.222 times 240,000 feet (see figure 1.1). A typical observing station at this latitude would then have employed 365 units of 24 inches in its circumference, scaling down 240,000 feet to 24 inches. Each inch would then equate to 10,000 feet (or 2 miles of 5,000 feet) on the parallel of latitude 48.189 north. One of the 365 day-units of 24 inches on the observatory circumference would equal 48 miles of 5,000 feet on the parallel, the equivalent of 72 miles of 5,000 feet on the equator.

To accurately arrive at a latitude like 48.189 degrees sounds complicated without modern glass and brass tools, vernier scales, bearings, and so on. However, this latitude can be found by viewing the circular perimeter of the observatory as both the equator and the meridian of a spherical Earth. The scientists of the megalithic, perhaps at Brittany searching for a more mathematically simple latitude, would have used the cosines of simple right-angle geometry to find a latitude whose cosine relative to the equator was a rational fraction, enabling whole number scaling of days on a new observatory circle at that latitude. The latitude of 48.189 degrees, with a cosine and thus parallel radius of two-thirds of the equatorial radius, provided a geometry useful for their work.

Finding the average altitude of a circumpolar star, that is, the point in-between the star's zenith and its nadir, would have given megalithic astronomers a fairly accurate altitude for the north pole. At this time the star Thuban was still approaching the north pole and, as we have said, a pole star is both rare and, at the best of times, an inaccurate guide to the pole's actual location. Instead, once an important site like Carnac was established, a list of circumpolar stars and their culminating angle directly north above the pole came to be known.

The culmination angle of a circumpolar star is the latitude of the site plus the co-declination of the star, at that epoch. For example, at Carnac in 4000 BCE, the extreme azimuth of alpha Ursa Major, today called Dubhe, was 40.1 degrees east or west of north, while the extremes in altitude (when directly north) would be an altitude of only 21.85 to 73.34, that is 25.73 degrees below and above the north pole. The extreme altitudes of the same star, at latitude 48.189 degrees will bracket again, by the same amount, a north pole that is now higher in the northern sky. The new altitude of the north pole, and hence its latitude, will be exactly between the stars' new extremes in altitude. The angular extremes in altitude of known circumpolar stars to the north of a new site have an angular average that defines the new latitude angle and the north pole. Viewing must take place from a raised backsight above a flat area surrounding the foresight, a vertical pole (or menhir) of known height and distance from the backsight. When looking for a specific latitude a degree or two further north, the backsights for a given star and the north pole must be lowered so that the new extremes of the star are aligned within the frame of the new north pole. The invisible north pole is now at the angular mean of the reference star's extreme altitudes. Such an apparatus was suited to finding a new latitude by lowering the observing position toward the flat ground, whereupon the same angular relationship of circumpolar culminations and nadirs confirms that a given latitude has been achieved.

Thus it was possible for the astronomers at Carnac to prepare a foresight that would work at Carnac but could also indicate when the equipment was at the correct new latitude. It might have only been used for extreme altitudes of a few circumpolar stars. The backsight would move lower than when it was at Carnac and only at the right latitude would circumpolar stars appear at the same angle above the top of the foresight as they had done at Carnac.

Having achieved the right latitude for a parallel two thirds the length of the equator, the same equipment would become the heart of a fully functioning circumpolar observatory for tracking the rotation of the earth from the new latitude of 48.189 degrees. The same apparatus, sent on an accurate bearing east or west, could travel many miles before being set up. Both observatories would need to maintain day-inch counting to predict lunar eclipses so that, when an eclipse actually started, the location of the marker star upon the circumference could be noted as a distance around the circle counterclockwise from a fixed point, probably north.

Each observatory circle was a model of the parallel of latitude, as it rotates with the earth. The difference in star placement on observatory circles between two observatories on the same parallel was the distance on the parallel between the observatories. This would have been a known distance, measured during the journey to set up the second observatory, and appropriately called the differential longitude. Knowing the distance between the observatories, the length (or circumference) of the whole parallel could be known. Also, having chosen the parallel of latitude that is two-thirds the length of the equator, the circumference of the equator would then be the inverse, three halves (1.5) of the circumference of the parallel.

If the same marker star and observatory circle were involved as that found at Le Menec's western cromlech, then we know that each inch on the circumference would equal 10,000 feet or two miles of 5,000 feet. For megalithic astronomers their two observatories might be 36 miles apart and they should therefore find a differential length in their eclipse observations of about 18 inches. It might have taken a century or two of improvements in sighting, widening of sites, and the creation of a network of sites to achieve their estimate for the equatorial circumference of 72 miles per day of angular rotation of the Earth.

We can now see that the measurement of the equator as 72 miles (360,000 feet or 4,320,000 inches) times the solar year came from the need to use the circumpolar observatory to observe the equator's size and from the observatory's circumference providing a scale model of the parallel of latitude on which the observatory sits. The parallel of latitude, as a small circle of constant latitude, is then of a length related to the equatorial great circle's length by the cosine of latitude. Choosing a latitude in which the cosine of latitude would be a rational fraction enabled the scaling of days on the observatory circle from and to their equivalent equatorial lengths to be rational as well. As we have seen in chapter 3, this scaling approach, so suited to metrological geometries, was seen within the dolmen of Mane Lud, where a scaled plan of the Locmariaquer complex was rendered.

THE METROLOGICAL SCIENCES: COUNTING, GEOMETRY, AND SCALING

We have proposed from the outset that megalithic astronomers did not always use what we see now but rather that the monuments we see are often permanent structures made to record their work done with less permanent media. Large stone structures with geometrical forms and long linear counts were not ideal for achieving the work that these instead represent. This is particularly clear with the long day-inch counts. Three solar or lunar years were soon exceeded when "counting" the 3,400 days of just half the moon's nodal cycle. Though count lines presented astronomical cycles in iconic form, human activity needed a reliable technology of counting that no longer relied on slavishly counting along geometrical

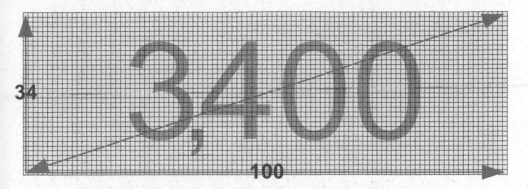

Figure 5.3. The large number 3,400 represented as a rectangular number through its factors of 34 and 100.

lines. It is far more relevant to break down the number into two meaningful factors so as to form a rectangle, with factors (side lengths) chosen to reflect a significant component of the periodicity of the number to be recorded. Taking the example of 3,400 days, the non-zero part carries with it the prime number 17, doubled to 34. Therefore one can make a measure 34 inches long and use it to define a vertical count that can be repeated 100 times on a flat surface. It is possible to make holes at each point, but in figure 5.3 the resulting rectangular number is shown as a set of small squares, each a notional inch square.

This approach is far more reliable to operate when repeatedly counting and thus represented a practical technology. It also demonstrates a developing interest in the use of multiple dimensions through the use of square and rectangular numbers, areas, and, ultimately, volumes.

The foot of twelve inches can be seen as an inevitable outcome of counting, since the inch is the width of the thumb and the fingers of the hand have twelve sections total. Counting on the fingers would have been the natural medium for human counting in the Stone Age,* and one simple technique is to count with the thumb the small bones of the fingers.

---

*Wikipedia for *Prehistoric Numerals* says "Counting in prehistory was first assisted by using body parts, primarily the fingers. This is reflected in the etymology of certain number names, such as in the names of ten and hundred in the Proto-Indo-European numerals, both containing the root d´k also seen in the word for "finger" (Latin digitus, cognate to English toe)."

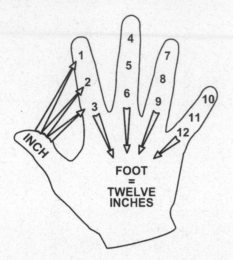

Figure 5.4. The twelve inches of a foot arose through counting on the fingers, where the width of the thumb is one inch and the fingers have twelve easily identifiable sections. We also show the thumb as the inch standard as well as the twelve counted inches in one standard foot. Adapted from Georges Ifrah, *The Universal History of Numbers.*

We can find further examples of hand counting techniques in historical times, some of which use the sections of the finger and thumb. For example, an English monk called "the venerable" Bede (672–735) proposed counting the 19-year Metonic cycle on a single hand and the 28-year solar cycle on both, while Muslims counted 33 on both hands three times to represent the 99 "incomparable attributes" of Allah (see figure 5.5).[1] We also find 33 at Locmariaquer, where it represents the solar hero cycle.

After developing practical technologies to handle the counting of very large numbers, the scientists of the megalithic developed methods of scaling so large cycles and measures could be represented in much smaller ways. Scaling seems at first a complicated method of calculation, something used in modern architectural plans. Before the computer however, draughtsmen used scaled rulers, which took real dimensions and immediately offered a scaled down length. The original measure is simply divided by a constant amount.

One simple non-calculating method of scaling used by the builders of the megalithic was to change the units of the original measurement, scaling down by using a subunit of the original unit, such as changing from feet to inches or years to months or day-inches. This was done at the dolmen of Mane Lud where a scaled plan of the full Locmariaquer complex was made, now hidden within the design of the Mane Lud dolmen. The scaling factor used by the megalithic designers was a unit of 40

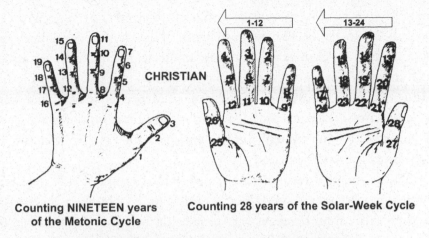

CHRISTIAN

1-12   13-24

Counting NINETEEN years   Counting 28 years of the Solar-Week Cycle
of the Metonic Cycle

Counting 33 of the 99 Incomparable Attributes of Allah

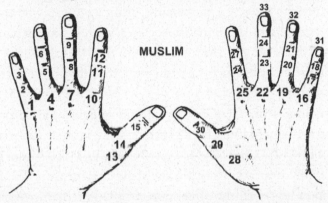

MUSLIM

Figure 5.5. Techniques for counting with the elements of the hand were well developed by the current era but probably originated long before as a way to envision aggregate numbers or measures such as the number of years in a cycle. Top left: Using the hand to count the Metonic cycle of 19 years. Top right: Using two hands to count the 28 years of the day-of-the-week repeat cycle. Bottom: The Muslim hand-counting method using two hands to count to 33, one-third of Allah's ninety nine names, can also be used to count the solar hero cycle of 33 years. Adapted from Georges Ifrah, *The Universal History of Numbers.*

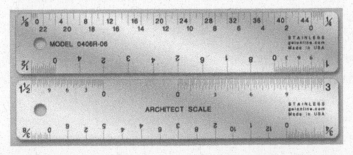

Figure 5.6. A modern example of scaled rulers used for making architectural plans.

megalithic inches to a megalithic yard so as to avoid the fractional division of 32⅝ inches or 33 inches per "yard." This allowed the dolmen to hold the plan for the whole monument at a scale of 1:40, one megalithic inch translating to 40 megalithic inches in the full-sized monument, indicating that both types of inch present at Le Menec (chapter 4) had two different uses: (1) an easy scaling factor of forty between the Mane Lud dolmen and the whole complex of Locmariaquer and (2) a day-inch count capturing astronomical periods.

Linear scaling was further developed in the organization of the later metrology where the English foot became a unitary measure for varied feet. In this, any fraction of that foot could be produced by a right-angled triangle to simultaneously multiply the foot by a denominator and also divide the foot by a numerator, using aggregates of any measure. This scaling enabled rationality between a circle's diameter and its circumference. For example, the foot of 35/32 feet, identified at Carnac, enabled a diameter of 48 such feet to equal 165 (5 × 33) English feet along the circumference.

Other scaling units were developed and widely used throughout megalithic times, including the Egyptian royal foot of 8/7 feet, the Sumerian foot of 12/11 feet, and the Saxon foot of 11/10 feet as well as very practical microvariations within the same foot measure. While useful in corresponding with the dimensions of the Earth, microvariations could provide rational whole units in both diameter and circumference by slightly varying the units of length, as we shall see in later chapters. Large aggregated units, like stadia of 500, 600, or 660 and miles of 5,000 or 5,280 feet were particularly useful for small-scale rendering within monuments of geodetic lengths, where the aggregate defines the unit for the scale. This is seen, for example at Edinburgh's full Royal Mile, Castle to Abbey, where the polar radius of 3,456 royal miles is represented by a single royal mile, a scaling of 3,456:1 with the polar radius.

Above all, metrology actively applied and maintained whole numbers, through different forms of scaling, to solve problems that would otherwise generate an irrational or decimal fraction. Creating ratios between units of length enabled any measurement to be transformed without breaking the implicit compact between *the-numbers-that-*

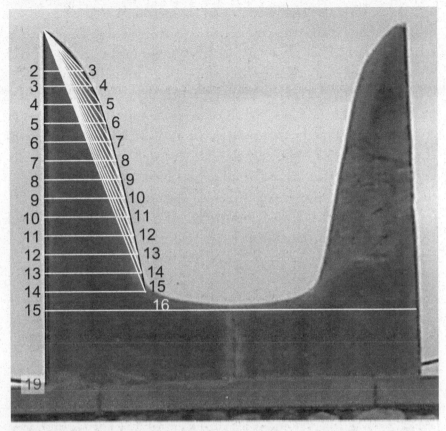

Figure 5.7. The Cretan "Horns of Consecration" (second millennium BCE) at the palace/necropolis of Knossos. Each horn may have been informed by a stone device whose curve gave pure number ratios of a standard foot through triangles, here superimposed upon it and whose cosines were all the simple harmonic ratios of N:N+1.

created-the-world and *the-project-of-understanding-the-world*. It was this compact that started breaking down in the ancient Near East, as writing could notate arithmetical numbers capable of transformation without the medium of geometrical lengths.

Geometry is the natural representation of scale and proportion in space. When a right triangle compares the lunar and solar year to find a four-square rectangle or when the triple-square approximates relations between the eclipse year, solar year, and thirteen month year, these years were being *made,* not by man, to follow a pattern involving small numbers that apparently had constrained these phenomena to relate to one another *in this way.* The value system for metrology that emerged was of calculations remaining rational so that the practitioner could stay related to the bedrock of number itself, reflected in the scaling between whole number measurements and employing rational fractions. This was glimpsed by Alexander Thom, who thought megalithic builders had, for instance, made pi equal three between some stone circle perimeters and their radius by flattening the original circle used to plan them. If megalithic people could only compute through counting, geometrizing, and scaling, then success in solving problems this way suggested to them that Reality also worked this way, opening the door to the widespread concept of a Creator, God the Maker, *Deus Faber,* or the Greek *Demiurge* found in the ancient world and who Plato proclaimed was "always geometrizing."

The different types and variations of the foot were formulated so as to scale and relate. The superparticular right triangle, whose sides differ by one unit (as in figure 5.7 on p. 125), enabled a different number of stan-

Figure 5.8. The Cairn of Gavrinis, c. 3500 BCE, found on the island of the same name in the Gulf of Morbihan, 4 kilometers east of Locmariaquer. Within it are 28 stones, most fully decorated with engraved designs of a unique character, all probably reset from other sites to the west and eventually sealed within as the megalithic period at Carnac was ending around 3100–3000 BCE.

dard feet on the base and the hypotenuse (like a cosine) to translate from one type of foot to another via a simple process of "vertical" mappings, from one side's metric to the other side's. The whole toolkit for metrology could thus be generated anywhere with a single example of the system, preferably the English foot, which then equaled the number one for the system, the unitary unit. That this happened was quite clear to John Neal who, from the structure of historical metrology, found different rational fractions of the standard foot underlying every historical length ever found.[2] All over the ancient world and even the new world, variations of the foot are found in use or clearly marked within monuments.

For the people of the megalithic, rigorous use of scaling reduced something impossibly complex into something of great mental simplicity, and in that respect the world in which they lived was thereby empowered. This only occurred through seeing the common unit between two time cycles, as at Le Manio where, between the sun and the moon, the difference in yearly day-inch count lengths is 10⅞ days.

Dividing these counted lengths by this difference gives their unique superparticular (N:N+1) signature ratio of 32.6:33.6. Three times their unit of difference of 10⅛ gives the megalithic yard of 32⅝ day-inches, a unit that would open many doors.

THE ROYAL CUBIT AT GAVRINIS

We have seen that Le Manio's Quadrilateral memorialized an early experiment in counting time using inches to represent passing days, the fulfillment of many time-factored objects produced over tens of thousands of years in the Upper Paleolithic era. At Locmariaquer and at Le Menec, inches are also evident in the time factoring of monuments. At Gavrinis, we find the key to the transformation from this inch-based form of astronomical metrology into the metrology inherited by the ancient world, a metrology based upon feet of twelve inches, rational fractions, and sophisticated scaling based upon harmonic ratios, the geometry of circles, and the modeling of the Earth itself.

The design of Gavrinis allows the sun to shine into the chamber only at the exact midwinter extreme of the sunrise to the southeast and only onto the left-hand end-stone within the chamber. The rays of the sun would be traveling down the now-familiar alignment of the FIVE side of a 3-4-5 triangle from the right-hand jamb of the entrance, past a guard stone, and into the chamber. The end-stone upon which the midwinter sun shines is engraved with many important measures (see figure 5.10 on p. 131). The central groove that rises from the middle of the stone and terminates at the point where the stone's side ridges join near the top is exactly 12/7 feet, the Egyptian royal cubit. The axis of the passageway tells a different story, being aligned to the southern extreme of moonrise during the moon's maximum standstill, which occurs once every 18.6 years or 6,800 days. The 28 stones that decorate the passageway and chamber within the cairn form a seven-square design relative to the passageway.

This seven-square design was fully intentional and connects important events of both sun and moon within the monument. As shown in figure 5.9, it is the seven-square that forms the angular bridge between the alignments of the sun's winter solstice sunrise and the moon's maxi-

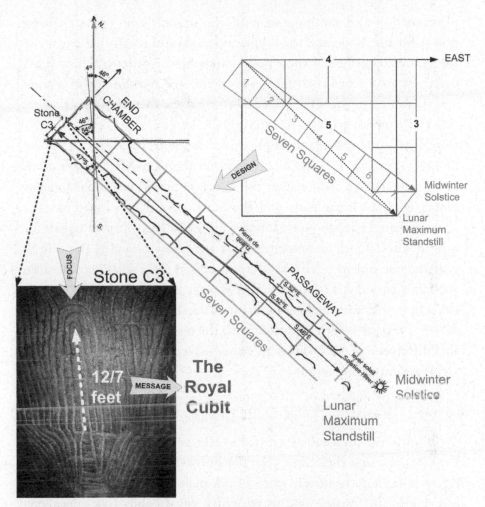

Figure 5.9. The architecture of the chamber and passageway within the Cairn of Gavrinis reconciles solar and lunar extremes through the relationship between the diagonals of the single-square, 3-4-5 triangle, and the seven-square. Stone C3, located in the chamber at the end of the passageway, marks where the sun would have shone at the midwinter solstice sunrise.

mum standstill moonrise—the diagonals of the three-by-four rectangle and the single-square, respectively.

So how did a 12/7 measure of the foot, the Egyptian royal cubit, become embedded in a monument employing the seven-square rectangle and having 28 (4 times 7) largely decorated stone panels? To answer this question requires a deeper reflection upon the eclipse phenomenon and the movement of the moon's nodes, the crossing points for the moon and the ecliptic. One should recall that when the sun stands upon one of these nodes, an eclipse is sure to follow. This occurs twice a year. The time between the sun standing upon a node and its return to the same node is 346.62 days, the eclipse year.

The simplest path to identifying the length of an eclipse year is to note that the dominant eclipse cycle, the Saros period of 223 lunar months, is *exactly* 12 lunar months shorter than the Metonic period of 19 solar years. We also see that the Metonic period is 7 lunar months longer than 12 lunar years and the Saros period is 5 lunar months shorter than 12 lunar years. Interestingly, the Saros period equals 19 eclipse years. Thus, one-nineteenth of the Saros period is the eclipse year and one-nineteenth of the Metonic period is the solar year. Because the Saros period is 12 lunar months shorter than the Metonic period, the eclipse year must be 12/19 lunar months shorter than the solar year. When the lunar month is equated with the megalithic yard of 19/7 feet, this difference of 12/19 lunar months between eclipse and solar years yields an interesting result:

$$^{12}/_{19} \text{ } Lunar \text{ } Months = {^{12}/_{19}} \times {^{19}/_{7}} \text{ } feet = \frac{12}{7} feet$$

That is, the royal cubit is created to represent the difference between the solar year and the eclipse year, but only when megalithic yards represent lunar months instead of 29.53 day-inches (or 0.75 meters). Without doubt, this was the means by which royal cubits first came into existence and the reason why stone C3 at Gavrinis has a royal cubit as its central motif. The stone clearly shows the relationships between the royal cubit and English foot and the whole stone appears framed within a circle with a radius of one megalithic yard (see figure 5.10).

Interpreting Stone C3

1. **The Sky Phenomenon:** The bottom left quarter of the stone represents how the lunar nodes in some years exceed the sun's solstice (lunar maximum) during each lunar orbit and in other years have a lesser extreme on the horizon (lunar minimum). The two features within dotted circles in this quadrant in figure

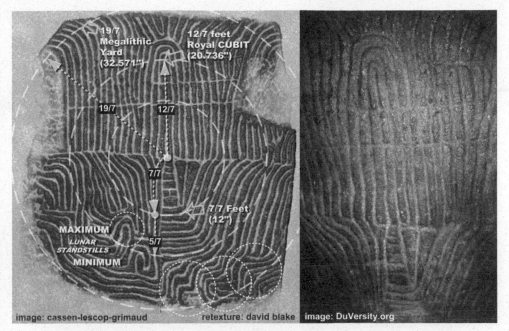

Figure 5.10. Stone C3 from Gavrinis appears to summarize the utility of the royal cubit and English foot if one uses the megalithic yard of 19/7 feet to represent lunar months rather than their length in day-inches. Base image generated by Laurent Lescop from 3D data scan, University of Nantes.

5.10 represent the maximum and minimum standstill moons, in the sky.

2. **Observing Horizon Events:** In the far bottom right area of the stone, indicated with dotted circles in figure 5.10, we see a representation of how, during the lunar maximum, the moon rises earlier than it would were it on the ecliptic and similarly, during the lunar minimum, the northernmost moonrise is later than it would be on the ecliptic.

3. **The Means Used:** The two downward-pointing triangles, in the lower right portion of the center of the stone, are each in the ratio 1:4, expressing the proportion of the four-square rectangle. These two arrow- or axe heads together form the angle of a triple-square and, to their right, carved lines like Le Menec's western alignments proceed at this angle. This angle is also expressed at Locmariaquer to the north and at the Le Menec to the east.

Both appear to involve the solar year to eclipse year ratio (of a triple-square).

4. **The Metrology or World of Measures:** The primary vertical, the central groove as noted earlier, is located centrally in the top half of the stone and measures 12/7 English feet, the length of the royal cubit. Drawing a circle upon the stone with this royal-cubit groove as the radius as well as one with an English-foot-length radius reveals interesting divisions made by the engraved lines, divisions akin to those found in a ruler. The downward-pointing radius of the English-foot circle coincides exactly with seven engraved units on the stone (the fifth and sixth of which are subdivided), and there are a further five units (the ninth and eleventh of which are subdivided) to reach the royal cubit's circle (see figure 5.10 on p. 131).

To the left of center in the royal cubit circle, there are 14 strips. This corresponds with the later Egyptian world in which there were 28 "digits" in a royal cubit. In this case there are 14 double digits, which may have been a precursor to the later 28 digits of Egypt. Examining the engraved divisions right of center, we find 19 units to the edge of the stone, which might have pointed to the 19 eclipse years of the Saros period, the 19 solar years of the Metonic period, or the 19 sevenths of an English foot in the astronomical megalithic yard.

The whole magic of this arrangement is founded upon the replacement of the 29.53 day-inch lunar month with the 19/7 foot megalithic yard, thus canceling the 7/19-month excess of solar over lunar year and the 12/19-month excess of solar over eclipse year. This replacement would have come about naturally once it was observed that the Saros and Metonic periods were commensurate in lunar months and were each made up of 19 years of differing types. A circle of radius one megalithic yard embraces the whole stone as if defining the field of this numerical achievement.

MEASURES BORN OF THE EARTH

When the Earth itself became the focus for megalithic interest, there was the question of how best to measure it. I have proposed that circumpolar observatories could have made the identification of latitude and relative longitude a possibility for the megalithic, enabling them to know where they were on the rotating surface of the Earth and how large that Earth was.

The relative longitude and distance between two observatories, directly east and west of each other, can reveal their angular difference in longitude through the angle of the circumpolar stars when recorded at a suitable shared moment, such as the onset of a predictable lunar eclipse. The difference in the star positions on the circumference of each observatory would then correspond to their distance of separation on their parallel of latitude, then knowable as a proportion of that parallel's total length.

Up to this point, angles were known through the geometrical structures that formed them (such as the angles of diagonals within multiple squares) or, as in celestial simulators, upon the movement of the sun or moon along a calibrated circle representing the ecliptic. In the same way the sun's cycle of 365 days meant that the circumpolar observatory's circle (representing the sidereal day relative to the sun's motion) could most naturally be divided into 365 parts. Today we do not use large circles to look at angles—because modern instruments such as the theodolite, diaoptra, and telescope are more accurate than the naked eye—this does not mean that large circles were not able to measure accurate angles when necessary.*

Using the number 365 on the circumference would have become a hindrance because of its factors, 5 and 73—metrological calculations made best use of smaller factors than 73. There must have been a move to the system of 360 degrees (with factors 2, 3, and 5) to define a complete circle. The notion of a degree of latitude evidently arose earlier than 3000 BCE because the geodetic metrology emanating from the megalithic made significant use of latitudinal degrees. This innovation of 360 degrees, credited to the Sumerians, must have been the work

---

*The limiting angular resolution of the human eye is about one minute of arc, which is 1/60 of a degree. On a large circle with 360 feet on the circumference, this amounts to 1/5 of an inch, when seen from the center and at Le Menec, 2/5 of an inch. Where results must be notated by setting a marker in place, one would not want, or need, more accuracy.

of whoever developed the metrology to divide the equator, using two circumpolar observatories, into 365¼ parts, each of 360,000 feet. This then meant, using the dual nature of this measurement, that there were also 360 degrees on the equator, each of 365,242 feet, numerically the solar year in days.

When 360 degrees were introduced to define latitudes, and hence the distance between them, the value of length of the degree in Brittany would have been found to be 364,670 feet, less than the 365,242 feet in a degree on the equator (and a lot more than the 360,000 feet per day in angle on the equator). The numerical similarity that arose between the number of feet within an equatorial day and the number of feet in a degree of latitude in Brittany implied a meaningful underlying proportionality governing the non-circular shape of the north-south meridian. This proportionality was represented in metrology by taking the number of feet in one degree of latitude and dividing it by 360,000. For example, if 364,670 feet were measured in a degree of latitude in Brittany, the foot for that latitude would be 364,670/360,000 or 1.01297 feet long and not far short of the Greek Geographical foot of 1.013476 feet, which divides the mean Earth degree found at Stonehenge.

Strong evidence has been found in our historical metrology for such proportionality in measures.[3] Many known historical measures divide into certain key degrees of latitude to give 360,000 feet, and most of the measures that have been preserved into recent centuries are rational variations of a definite type of foot, rational in two ways: to the English foot and to the dimensionality of the Earth. In other words, rational variations were made available in the families of ancient measures and made key latitudes divisible by a known whole number.

When one looks at the variation in degree lengths, one discovers that over a relatively wide range of latitude, from 25 degrees to 55 degrees, the degree lengths increase at a linear rate of about 63 feet per degree (see figure 5.11). This meant that the megalithic system would have been able to calculate the lengths of other degrees based upon the length of a degree at key latitudes: Ethiopia in Africa, Delphi in Greece, and Avebury in England. For example, around the latitude of Athens and Delphi, the degree length is 360,000 units of a Greek foot, 1.01146 English feet. This was a sophisticated ratio of the English foot, being

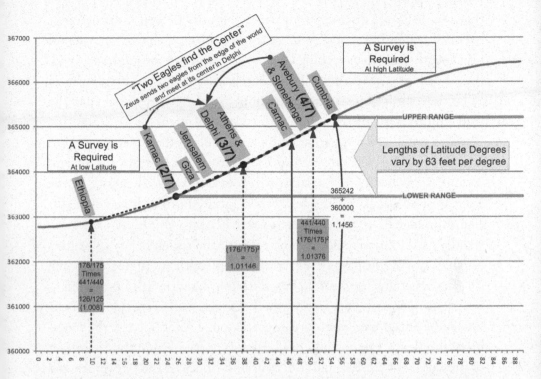

Figure 5.11. Modern measurements from the Smithsonian and Egyptian approximations from Livio Stecchini[4] show the variation in degree length having a constant rate of change between Karnak at 2/7 degrees north and Avebury at 4/7 degrees north. The Greek tale of Zeus sending eagles from either "edge of the world" to find the Omphalos "center" of it can be seen as identifying the oracular center of Delphi as being at the latitude 3/7 degrees north, between the megalithic and pyramid-building cultures who had measured the Earth's dimensions.

$(176/175)^2$ of a foot. Another key foot, the Geographical foot of the mean degree (51–52 degrees), is further increased from that at Delphi, so as to be 441/440 larger. These ratios will be explained later, but the point is that knowing which latitude they refer to allows you to extrapolate over the range 25 to 55 degrees north to obtain the length of another degree. Or, if you know another nearby degree length, one can add or subtract 63 feet times the difference in latitude. This would soon have become clear from consecutive degree lengths in that range, perhaps triggering a search for where this rule broke down in the Tropics.

It is obvious from this that the geometers of the megalithic moved to using 360 degrees in a circle. Their metrology would have found that the distances discovered for degree lengths fit well into the range between 360,000 feet and 365,242 feet, enabling the microvariation of unit lengths and then making some of these suited to rationally dividing into some key latitudes. These key latitudes enabled extrapolation in latitude while also standing as exemplars for the method of adapting your units so that they fit the length to be measured.

The complicated situation, of each degree length growing, was thus largely avoided in the range 25 to 55 degrees, that is, from the ancient Near East to northern England, by observing the differential in the lengths of consecutive latitudes. The associated problem was that this flat zone made the Earth's north-south shape—an ellipse called the geoid—hard to determine without a trip to the region below 25 degrees north. This would have necessitated a trip south to 10 degrees north for the Egyptians, who would come to dominate the gold produced at those latitudes, in Nubia and nearby Ethiopia, the latter famous for the Queen of Sheba who married King Solomon. Alternatively, the megalithic people in Britain could head further north beyond the 55 degrees of present day Cumbria and into Scotland or across the North Sea and into Denmark and Sweden, where strong megalithic traditions are also found to have existed.

MEDITATIONS UPON THE FORM OF THE EARTH

There would have been a number of ways of looking at the Earth even before trying to measure it and determine its size and shape. The first way, most simple yet quite accurate, is to visualize the Earth as spherical, that is, as a spinning ball whose equator and north-south meridians are perfect circles—the "mean Earth." This is the first stop in getting beyond the popular idea that people in the past thought the Earth was flat, just because the bit they were standing on looked pretty flat. The astronomers of the megalithic were evidently able to see not only the sun, moon, and stars as going around the Earth and the shadow cast upon the moon at eclipse as the Earth's and not a dragon but also that the moon and the sun were both balls and hence the Earth likewise was also a ball.

Figure 5.12. Left: The distribution of megaliths, in the form of standing stones or menhirs, matches the requirements to survey the north-south meridian for its latitudinal degree lengths. Right: Multiple-square geometry appears in the megalithic monuments of Scandinavia, such as this seven-square at Langdysser, Denmark, and this suggests links to an earlier Carnac where multiple-squares were developed as an early geometry for astronomy. Base images from Google Earth.

Joining up a day-inch count for four years created a solar simulator, making the ecliptic a circle around the Earth. The moon's orbit similarly made sense as traveling upon a circle with the Earth at the center. Once astronomical time on Earth was studied using day-inch counting, a geocentric or Earth-centered pattern of time, largely valid only for the surface of the Earth, was discovered. It was not relevant to this pattern of time whether the Earth went around the sun, as we know today. It was simply useful to know that time on Earth was some kind of a key to life on Earth.

When considering a spherical Earth, its rotation ever onward to the east causes everything in the sky to move continuously west, between the eastern and western horizons. This leads to the idea of standing on a north-south segment of the rotating Earth, "your bit" of the Earth in which the north never moves but everything else moves above you. These two ideas, of your segment (the meridian) and what it experiences every day, were perfectly portrayed in Mane Lud where two men were halved down their centerline, leaving only their right-hand sides. One was darkened and the other light, representing the segment that was their local segment of the Earth, dark by night and light by day. (You may remember, they were being pulled by posts with horse heads marking the extremes of the moon.)

Another good use of the spherical model is to see that one could build a cube around it in which two of its dimensions would be diameters of the spinning Earth, while the third would be the polar diameter. For a spherical Earth this would be a perfect cube with equal side lengths. However, when a planet starts spinning, the equator gets fatter, stealing mass from the interior, and, as a consequence, the poles of the Earth are slightly smaller than the equatorial radii. This leaves the cube model with two dimensions that have grown and one dimension (pole-to-pole) that has shrunk. The unspun Earth remains an important reality though, since its circumference, initially identical whatever direction one travels, becomes only the north-south meridian of the spinning Earth.

To summarize:

a) The spherical Earth, called the mean Earth, while not accurate to the shape of the Earth's meridian, gives the true length of the meridian of the actual Earth that spins.

b) The spinning Earth grew at the equator, making the equatorial radius greater than the mean Earth's radius.

c) The polar radius of the Earth shrank due to the Earth's spin, and through circumpolar astronomy the polar axis defined the cardinal directions north, east, south, and west.

d) Both the equatorial radius and the polar radius then fit as the major and minor axes of an ellipse whose perimeter is the same length as the circumference of the perfectly spherical mean Earth.

In the next chapter we will examine the philosophical and religious ramifications of this, which embraced many of the important themes for later civilizations, but the challenge for the geometers of the megalithic was to work out what these key radii and circumferences were. It is a task that from a modern standpoint would seem comically overoptimistic armed only with stone age tools, but, as with astronomy, the people of the megalithic would show themselves equal to the task, leaving the evidence within their metrology and monuments, these being their language and documents.

We saw earlier that some parts of megalithic metrology were devoted to rendering the lengths between parallels of latitude rational to specially adapted lengths of feet. Other aspects of their technology defined lengths that embraced the key dimensions of the Earth within a model of the Earth that cleverly used three rational approximations of pi, all to give facts in the simplest and hence most portable and memorable form.

This arrangement using three values of pi was made compatible with the variations of the foot to suit key latitudes through the fiendishly clever derivation of two ratios that were themselves partial products of two of these rational pi fractions.

Relating the Pole and Mean Earth

The first ratio is 441/440, which is the result of multiplying 63/10 and 7/44. In this ratio 63/10 as two pi enabled the polar radius to be accurately deduced from the mean Earth circumference (which, as stated above, is the meridian length for the actual Earth). The second ratio, being the inverse of the more accurate 44/7 value of two pi, enabled

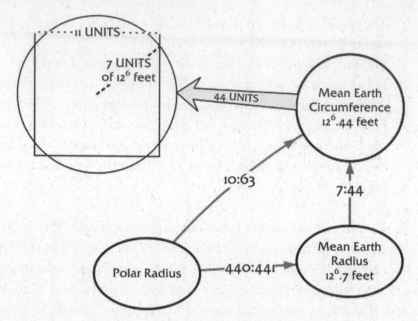

Figure 5.13. The model of the Earth in which the polar radius is made rational to the mean Earth using the differential ratio of two rational approximations to two pi.

the circumference of the mean Earth to be related to the mean Earth radius. The combined effect is to enable the polar radius to be rationally connected with the radius of the mean Earth and the mean Earth circumference, as shown in figure 5.13.

One can notice another "relic of the creation" in that the mean Earth appears close to being divisible in both its radius and circumference by twelve to the power of six so as to leave the relatively accurate two pi approximation of 44/7 exposed. In chapter 7 we will see how this relic will lead to a later interpretation of Y.H.W.H (6.5.10.5) as 6^5 times 10^5.

Relating to Latitudes and Cubes

The second differential ratio related to variations of the foot at key latitudes is 176/175, which is the product of 44/7 and 4/25. We have seen that the meridian on which latitudes are marked is rational according to 44/7 and that the key latitudes shown in figure 5.14 involve products of the ratios 441/440 and 176/175. The latter ratio had to be useful, for it is able to cancel 44/7 but also produce something more fundamental.

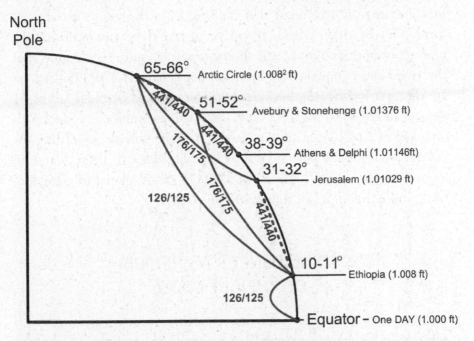

Figure 5.14. The dimensional expansion of the length of degrees between key latitudes, creating feet that divide into latitudes based upon one day in angle on the equator equaling 360,000 feet (after John Neal).

The primary use of 176/175 was its ability to create two versions of the same foot, which would make a radius defined in one foot equal a whole number of feet measured in the lesser foot (for a full statement, see page 167 in chapter 6).

It was found that these two measures had a further and unexpected benefit for ancient number science in that the product of 441/440 and 176/175 is 126/125 or 1.008 in decimal notation. This had a very important role in measuring volumes and working with cubes. Noted in ancient scriptures, the ratio 126/125 has the ability to double the volume of a cube, a task associated with the Greek god of the Earth, Apollo, whose altar was cubic. The best fractional approximation to doubling the cube is to multiply a cube's side lengths by 126/100 (5/4 × 126/125). This creates a rational side length, 126/100 times the original, and a volume just one five thousandth over a doubling.

Returning to the role of metrology in north-south surface

measurements, it was found that the lengths of degrees at specific latitudes vary by either one or the other of just these two ratios or the product of both as 126/125. It therefore seems, quite remarkably, that the larger ratio employed within metrology, namely 176/175, had yet another use in bridging between two pairs of key latitudes (10 to 51 degrees and 31 to 65 degrees), yet it also forms a combined product with 441/440 of 126/125, helping to double a cube's volume (exceeding the volume by only one part in 5,400). The latitudes 10 degrees and 65 degrees north contain some of the 126/125 ratios related to cube doubling and harmonics (see figure 5.14 on p. 141).

COINCIDENCES AS RELICS OF THE CREATION

As one ventures on into what the megalithic created and how they achieved it, the more our words to excuse their creativity as a coincidence are revealed to be acts of euphemism. This is illustrated when, through the technique of day-inch counting, a new unit emerged at Le Manio called the megalithic yard, this being an early effort to relate the relative difference in duration of the solar year and the lunar year (of twelve lunar months). The inch, if based on our thumb, would seem an arbitrary though convenient choice of unit for the task. It "made" a megalithic yard larger or smaller, according to its length. Then, using that length differently, to count months as megalithic yards, the difference between the eclipse year and the solar year reveals the royal cubit of 12/7 feet, and so this too should be a purely relative measure.

But as we shall see, the dimensions of the Earth, surely not themselves arbitrary, have many developed relationships to these supposedly arbitrary measures of astronomical time found in Carnac, revealing a pure relativity between these celestial bodies. But this is not the case, for within the Earth-sun-moon-planets manifold lies a deeper developmental history. If we reduce the royal cubit to its royal foot of 8/7 feet, we obtain a polar radius of 3,456 royal miles. It is therefore true that the difference between the eclipse year and solar year (12/7 feet, the royal cubit) relative to the difference between a single lunar year and single solar year (the English foot of twelve inches), counted in megalithic yards, relates to the Earth's polar radius. However, this megalithic yard is actually the difference between 19 lunar years and 19 solar years, when counted in inches that have been *slightly expanded* (by 32/29) to create the excess of the solar month (30.43 days) over the lunar month (29.53 days). Confusion arises when attempting to grasp this, a confusion proportional to the apparent intelligence required to achieve such coincidences.

It therefore appears impossible to separate the Earth from the moon and the sun because these are all, in fact, part of an interconnected system, one that our culture believes are only material aggregations. Another factor in this planetary system is often missed: the human factor. If we include the planets as integral factors alongside the Earth, moon, and sun, then human beings are a fifth part in this system: a quintessence within which this drama now acts itself out before an observer who can understand. The inch that begat the megalithic yard, then being slightly enlarged to embrace the excess of the solar month over the lunar month, connects us to the tradition of *twelveness* as the autocratic ruler of the planetary system. But we need to see the human factor not as a saber rattling of Zeus on Mount Olympus but as something small that can yet have a great effect.

The inch connects us to the figure of Tom Thumb who is more than a storybook character. Tom Thumb portrays the relationship to a traditional miniature man, the size of a thumb, who was influential in defining, as a star, the celestial meridian for a precessional age. He was associated in an unlikely way with the seven stars of Ursa Major, also called the seven sages, and the raven who carries him off is the traditional symbol for Saturn. As de Santillana and von Dechend tell us in *Hamlet's Mill,*

> The Rishi Vashishta is unmistakably zeta Ursa Majoris whilst Alcor, the tiny star near zeta (Tom Thumb, in Babylonia the "fox"-star) . . . This is the "birth" of the valid representatives of the [earth's] poles Mitra and Varuna" whose seed gave [Vashishta and Tom Thumb] birth.[5]

Von Dechend goes on to infer that Mithra is the "ship's belt," which would then be the celestial equator. In his *A Classical Dictionary of Hindu Mythology and Religion,* John Dowson gives Varuna as

The universal encompasser, the all-embracer . . . One of the oldest of the Vedic deities, a personification of the all investing sky, the maker and upholder of heaven and earth. As such he is the king of the universe, king of gods, and men, possessor of illimitable knowledge . . . He is often associated with Mitra, [Varuna] being the ruler of the night and Mitra of the day.[6]

This seems to give clues to the two halflings incarcerated at Mane Lud in a chariot pulled by lunar maxima.

Varuna is associated with counting in days as inches. From the Rig Veda:

This great feat of the famous Asurian Varuna I shall proclaim who, standing in the air, using the Sun as an inch scale, measured the earth.[7]

The Asura, like the Titans, were prior to the sort of gods found on Mount Olympus. They are associated with a golden or previous better age and therefore with the structure of the world and a knowledge of the precession of its equinoxes, for what lies behind the light of a constant solar year is an ever changing backdrop of stars in each season's night sky. Kronos, who parallels Varuna in ruling over a golden age, holds further parts of the picture. The 13th Orphic Hymn to Kronos addresses the god as,

> Father of the blessed gods as well as of man, you of
>   changeful counsel, . . . strong Titan who devours all
>   and begets it anew [lit. "you who consume all and
> increase it contrariwise yourself"], you who hold the
>   indestructible bond
>   according to the apeirona (unlimited) order of Aion,
>   Kronos father of all,
> wily-minded Kronos, offspring of Gaia and starry
>   Ouranos . . . venerable Prometheus.[8]

There is a great deal to know within this "data compression," but von Dechend suggests we use a lookup table where "bond" means "inch scale" and "to found" means "to survey" (indeed, she did that for us with the Rig Veda quote above).

> If instead he were to read "inch scale" and "to survey"—a divine foundation is every time a "temenos" [Gk for sacred precinct]— [one] would promptly react in a different manner. Kronos-Saturn has been and remains the one who owns the "inch scale," who gives the measures, continuously, because he is "the originator of times," as Macrobius says, although the poor man mistakes him for the sun for this very reason [note: *Sat.1.22.8: Saturnus ipse, qui auctor est temporum.*]. But "Helios the Titan" is not Apollo, quite explicitly.[9]

We shall have to come back to all this, but the principles are that the inch is something employed by the God of a certain type found in many different mythological systems and Tom Thumb appears to be a symbol then associated with the polar regions and the precession of the equinoxes. Second, this leads us to the notion of "a survey" of the Earth, which is how this divine system came to be known, a system too impossible in its perfectly interlocking parts to be a human creation or a coincidence. "The world" should not be mistaken as just being all of its necessary parts but rather as being an *integrated whole* with which early efforts to understand "the world" came into contact with a pattern beyond their wildest or rudest imaginations. Carnac was just the beginning in a developing vision— from Proclus' commentary on Plato's *Cratylus*:

> The greatest Kronos is giving from above the principles of intelligibility to the Demiurge [Zeus], and he presides over the whole "creation" [*demiourgia*]. That is why Zeus calls him "Demon" according to Orpheus, saying: "Set in motion our genus, excellent Demon!" And Kronos seems to have with him the highest causes of junctions and separations . . . he has become the cause of the continuation of begetting and propagation and the head of the whole genus of Titans from which originates the division of beings [*diairesis ton onton*].[10]

It is these "principles of intelligibility" that enabled the megalithic geometers to understand this world through their right choice of *measures native to the creation,* in both astronomy and in their survey of the Earth, first in England and then in Egypt.

Today we are left with poetry where once there was a prose, of discoveries that became noted as legend and myth, whose oblique allusions protect both the gods and their messengers. To cure our lack of a history for the megalithic period, we must create one using whatever channels are available, including the human imagination. I will take an early step by saying that two successive and proximate cultures, which did similar things in Brittany and Britain, were the same people. They were capable of moving whenever and wherever they needed to in order to continue their definite trajectory of learning about the world in which they lived. They developed an exact science that could have been the science of the gods themselves, and from their work would come all that we know about the gods within the religious sentiment of wishing to connect oneself with the source of existence.

> *And I saw in those days how long cords were given to those angels, and they took to themselves wings and flew, and they went towards the north.*
> *And I asked the angel, saying unto him: "Why have those (angels) taken these cords and gone off?" And he said unto me: "They have gone to measure."*
> *And these measures shall reveal all the secrets of the depths of the earth . . .*
>
> THE BOOK OF ENOCH, CHAPTER LXI,
> VERSES 1, 2, AND 5[11]

6

THE MEGALITHIC
MODEL OF THE EARTH

THE MOST NATURAL approach in studying the Earth as a three dimensional object is as a perfect sphere. This means that an observatory circle, scaled to its parallel of latitude and to the equator, was a ready-made representation of the Earth both as the equator and as the meridian. Viewing the observatory circle as the Earth's meridian would have enabled the geometry of the circle to give access to the circumferential length of different parallels of latitude by building right triangles relative to the center of the Earth. This method assumes a geocentric idea of latitude rather than one related purely to the stars. Once the length of the equator was known, the geocentric latitude could be calculated through the length of the parallel relative to the equator, measured in the usual way but now revealing itself within geometries that could be constructed within the observatory.

This way of measuring latitude, from estimating the circular length of a parallel of latitude on a spherical Earth, differs little in potential accuracy from measuring astronomical latitude and it had a key advantage for megalithic scientists. Establishing geocentric latitude did not rely on equipment pointing at the sky to measure angles on the horizon. Megalithic scientists could not "take measurements" as we do from a device (such as an instrument pointed at the sky) because they had no notation

suitable to abstract measurements except as lengths. There always had to
be a metrological means to store data and the large circles of observatories
could deliver accurate angular measurements approaching the one minute
of arc possible to the naked eye. Measuring latitude within a geometrical circle brought the measurement of the length of a parallel of latitude
within the scope of triangular mechanisms, which could reveal angles.
The megalithic form of notation for an angle was always the tangent (the
drop angle) or the cosine, realized within a right triangle.

The example of Stonehenge in figure 6.1 shows the way in which
geocentric latitude would be presented using the circular geometry of a

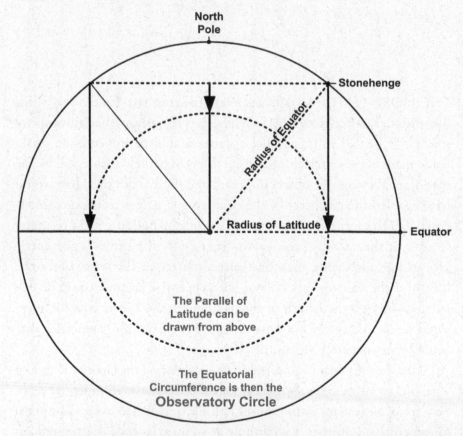

Figure 6.1. The measurement of geocentric latitude by comparing the length of a
parallel of latitude with the length of the equator. The observatory circle used for
measuring a parallel can be reused to establish the latitude of that parallel relative to
the known circumference and radius of the equator.

geodetic observatory as well as how the length of the parallel at Stonehenge could be deduced. The radius of the parallel can be directly compared as a ratio to the radius established for the equator itself. One can form a right triangle where the hypotenuse is the equatorial radius and the base the radius of the parallel of latitude. This type of geodetic latitude measurement avoids sighting the north pole and the use of a pole star, which as we've previously discussed is a very rare phenomenon (see figure 6.2).

In figure 6.2 one can notice that the triangle is the same as the geometry within the geodetic observatory, where the dotted curve is on the observatory's circumference and represents a part of the equator or of the meridian, assuming a spherical Earth. The radius of the parallel is 12 units and the radius of the equator 19 units, relating to the royal cubit of 12/7 feet and the astronomical megalithic yard of about 19/7 feet (the discovery of this unit to be discussed shortly). One English

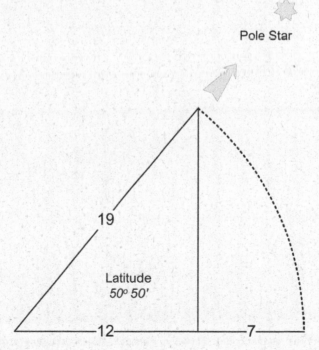

Pole Star

19

Latitude
50° 50'

12   7

Figure 6.2. The classic view of determining latitude (near Stonehenge) from the north pole using a pole star, which ignores the fact that pole stars are very rare and never exact enough to determine latitude accurately.

foot is 7/19 of this megalithic yard, being the 0.368 of a month by which the solar year exceeds the lunar year, each year. There is something about the placement of Stonehenge that relates the Earth to the sun and moon. The key to choosing its location appears to have been the island appropriately called Lundy on the outer limits of the Bristol Channel in England (see figure 6.3).

Stonehenge is located some 240,000 astronomical megalithic yards (AMY) east of Lundy, on the same parallel of astronomical latitude, 51 and 1/6 degrees. In 3100 BCE Stonehenge consisted only of a circular bank and ditch about 360 feet in diameter ("Stonehenge 1") with a

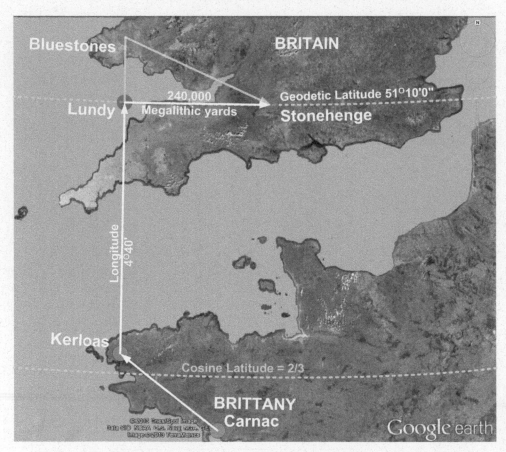

Figure 6.3. The geodetic link between Europe's largest standing stone, Kerloas in western Brittany, and Lundy Island, which is on the same latitude as Stonehenge and the same longitude as the site where the bluestones of Stonehenge originated. Base image from Google Earth.

largely clear space within and probably no stone circles. Within the circular bank and ditch were 56 "Aubrey holes," arranged in a circle about 283.3 feet in diameter. Cut into the chalk, they could have been holes for wooden posts (like tree trunks) or for the bluestones sourced from a district in Wales 100,000 AMY north of Lundy Island. Lundy, the bluestone source site, and Stonehenge form an apparently intentional landscape geometry, a 5-12-13 Pythagorean triangle (see figure 6.3). The circle of holes was given a diameter of 104 (8 times 13) megalithic yards. Some twentieth-century interpreters thought the Aubrey Circle was used for simultaneous simulation of the sun, moon, and eclipse nodes, using the number 56, rather than the 82 units that could only simulate the moon.[1] In a geodetic observatory 56 could also have signified the septenary division of latitude common in the ancient world.

The diameter of the Aubrey hole circle is just short of 3,400 inches, suitable to count the lunar node (eclipse) cycle of 6,800 day-inches across the flat area of Stonehenge 1 (the same cycle recorded in megalithic inches across the western cromlech of Le Menec). To measure their relative longitude, two geodetic observatories would need to track the eclipse nodes relative to the sun and moon so as to anticipate those important

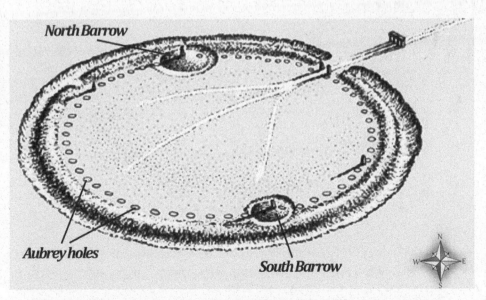

Figure 6.4. Stonehenge I, which included the circle of 56 Aubrey holes and the two mounds on opposite sides of the circle's diameter. Illustration from English Heritage.

moments when the shadow of the Earth first appears upon the full moon during a lunar eclipse. As discussed previously, lunar eclipses offered the perfect opportunity to synchronize two observatories and initiate a measurement at each for comparison with the other site's own measurement, thus calculating the relative longitude between the two.

If Stonehenge was one of two identical observatories, then the other was probably on Lundy Island, which at the time was a hill on a coastal

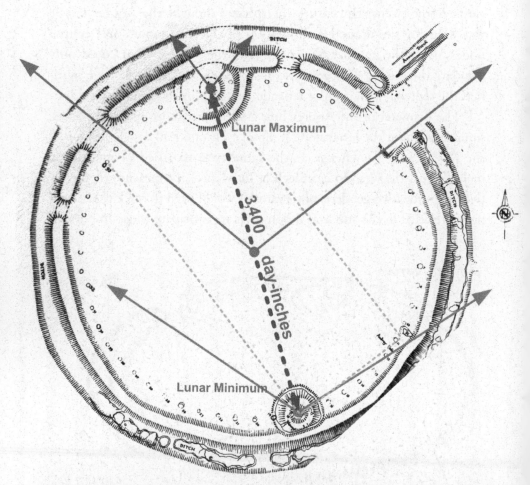

Figure 6.5. The layout of Stonehenge I, within the ditch, bank, and Aubrey Circle, suggests an area within which a circumpolar observatory could operate. On the same latitude two such observatories could track the sun and lunar nodes, using a count of 3,400 day-inches and a 56-hole simulation of the sun and moon, to predict lunar eclipses and thus measure differential longitude. Taken from the original survey plan, courtesy of A. S. Thom.

plain in the process of being inundated by rising sea levels. By accurately defining a fixed distance of 240,000 megalithic yards between Lundy and Stonehenge, megalithic geometers could obtain an accurate measurement of relative longitude and thus the length of that parallel of latitude as well as their geodetic latitude. It is this that may account for the less-than-perfect location of Stonehenge upon a slightly sloping site.

The eastern observatory (Stonehenge) would be 2° 50' 46" more advanced (clockwise) than the western observatory (Lundy). The angular difference (about one 126th of a complete circle) means that the length of the entire parallel should be 240,000 AMY times 126.488, or 15,612.28 miles. This compares to an equatorial circumference of 24,902.86 miles in the ratio is 0.626927, the cosine of Stonehenge's latitude of 51° 10' 35", just seven seconds less than its present value, 51° 10' 42.2" (the difference results from using a geocentric [or *reduced*] latitude, which is based on a spherical Earth). It was impossible in the megalithic era to obtain such an exact result from the alternative sky-pointing technique.

Since the equatorial radius was a known proportion of the measured length of a parallel, then the converse was true: that anywhere on the planet, an estimate for the length of a parallel of latitude would give the geodetic latitude, once one knew the length of the equator. It is amazing that ten seconds of accuracy might have been possible in megalithic latitude measurements and also how the means used provided such a powerful visualization of the Earth and its other latitudes. Any large circle when calibrated could be a model of the Earth.

THE GEODETIC KEY TO STONEHENGE

If megalithic achievements were the result of *possible* acts, then we are not faced with the need for a preceding super-culture to explain what the megalithic scientists were doing and how they were doing it. We now come to the central question for this chapter. Is it at all possible for somebody on the surface of this planet to have developed a model of the Earth? How could the polar radius have been deduced without surveying the most curved parts of the Earth's crust near the equator, below ten degrees north?

Metrology appears to have developed from primitive aggregates, sufficient and necessary to achieve the astronomical tasks of Carnac's monuments. The demonstrable tools of day-inch counting and circle building for geodetic observation have identified specific units as used in their construction, such as the inch, the foot (12"), the month (29.53"), the megalithic yard (32.625"), the vara (33"), and the none-too-recent meter (39.37"*). The metrological ratios of historical metrology cannot be assumed as having come into use in Britain until they were needed or to explain the dimensions found within actual monuments. They were clearly tied to counts of astronomical cycles of time and the sacred measures of the Earth and were developed well before the construction of Stonehenge.

The end-stone of Gavrinis displayed an astronomically valid royal cubit (12/7 feet) and by implication therefore an astronomical megalithic yard (AMY) of just over 19/7 feet. These septenary units, with a denominator of seven, perhaps indicate how fractional units of a foot came to the notice of megalithic astronomers as a useful tool for expressing numerical transformations of scale. The AMY would certainly prove crucial to what was achievable in Britain. The megalithic astronomers must have improved on the megalithic yard of Le Manio, calculated over three years and not fully true to the astronomical ratio between the solar and lunar years. Only the more accurate anniversary of sun and moon within 19 years can provide the unit called the astronomical megalithic yard, and this could have been measured at Locmariaquer, see figure 6.6.

The astronomical megalithic yard (AMY) is calculated thus: 19 solar years (of 365.2422 days each) minus 19 lunar years (of 354.367 days each) equals 206.62728 days, which is 6.997 lunar months (of 29.53059 days each). Dividing 19 lunar years by the excess 6.997 lunar months, it is seen that 2.715426588 *lunar* years are required for the excess to equal a single lunar month.

The length of the AMY is therefore 2.715426588 feet (in feet, of 12 inches, because there are 12 months in a lunar year). To put this into septenary units, we multiply by 7 and arrive at 19.00798612 or 19.008

---

*This is the standard adopted by the United States and appears to be 4/3 of the lunar month day-inch count found at Le Manio.

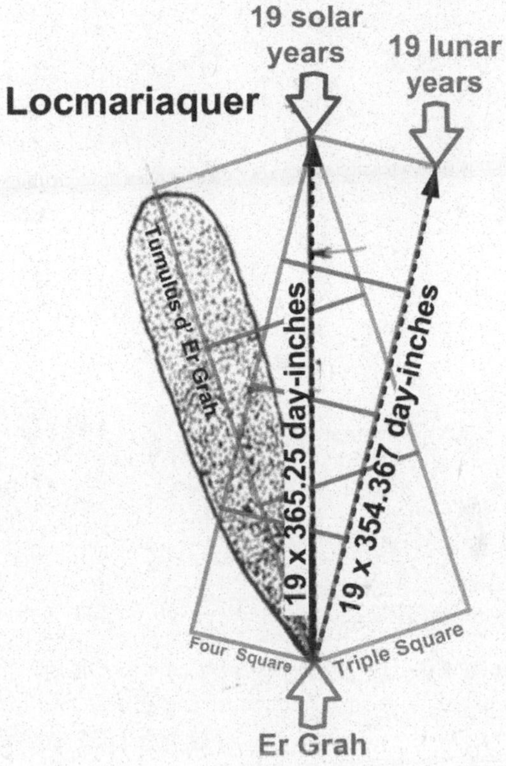

Figure 6.6. Practical derivation, using the long counts within Locmariaquer, of the astronomical megalithic yard from the 19-year Metonic period's excess over 19 lunar years.

to one part in 1,369,330. Therefore, for practical purposes, the AMY is exactly 19.008/7 feet long.

Since the megalithic astronomers could not use a decimal formula like 19.008/7 feet or a cumbersome 19,008/7,000, they would have sought a rational formula for this unit, which would later prove to be a great boon when used in Britain at the latitude of Stonehenge.

We have noted that megalithic astronomers and geometers naturally kept numerical factors separate from each other, multiplying them in what we could call a rectangular number, an area calculation. Without symbolic multiplication a repetition of the same length a number of times created the product when required. One notices that the AMY is a "royal" unit of 19 divided almost perfectly by 7. The attempt to find a rational AMY must have been tried, such as 18/7 of another type of foot.

This ratio, 18/7, is a royal version of a larger foot. If we multiply twice by three halves to make a cubit (3/2) squared and then multiply

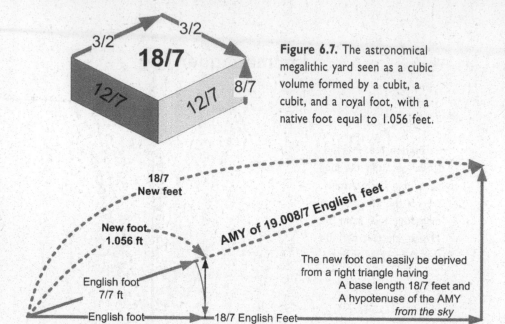

Figure 6.7. The astronomical megalithic yard seen as a cubic volume formed by a cubit, a cubit, and a royal foot, with a native foot equal to 1.056 feet.

Figure 6.8. The generation of a new foot length that turned the AMY from a then inexpressible decimal fraction of English feet into a rational fraction, of 18/7 of these new feet 1.056 feet long.

by the royal (8/7) foot, we form a cubic volume of 18/7 units as shown in figure 6.7.

This act was crucial as the new type of foot it exposed happens to be the geographical foot for latitude 51 to 52 degrees north, where Stonehenge and Avebury were built, one quarter of a degree apart. Discovering the new foot was reasonably straightforward through the use of English feet within a right triangle, as in figure 6.8.

The foot then formed was 1.056 feet long, and it would eventually be called the *root canonical Persian* foot, in today's (rather mongrel) terminology, partly "ancient" (*root canonical* after Neal[2]*) and partly "his-

---

*Neal's metrology, a resolution of John Michell's in *Ancient Metrology,* is firmly ignored by antiquarian professionals while historical metrology, like that found in the work of Berriman, is simply too complex for modern readers to recognize the interrelation of ancient units, which did exist. Neal's great breakthrough was when he saw that historical metrology contained an overall pattern, in which all units had a fractional relationship to the English foot, chosen to act as the number one for the system, its unitary unit.

torical" (*Persian* after where it was found[3]). We can and must ignore the metrological system that called this foot "Persian" because this system was not developed until hundreds of years after Stonehenge 1. For the megalithic geometers, there was only the need to see the AMY, a length derived from day-inch counting, as rational and as 18/7 of a new type of foot relative to the English foot of 12 inches. Recognizing this, we can see what happened in Britain in proper perspective, with no Atlanteans or aliens necessary to achieve the impossible.

The AMY must have been the seed for the eventual creation of the metrological system since its foot, of 1.056 feet, contains all the elements found in the later metrological system. First, it is a variation of a rational relative of the English foot, the ratio 21/20 (1.05) feet long. Second, 21/20 has been varied (not by man) by 176/175 and this was the first ratio used within metrology for microvariation. This ratio was extremely useful in obtaining a rational circumference from a rational radius measured in units that are 176/175 of the units of the circumference. Third, this new type of foot is related to another rational unit 25/24 (1.0416) of the English foot as follows:

$$25/24 \text{ feet times } 441/440 \text{ times } (176/175)^2 = 132/125 = 1.056 \text{ feet}$$

We also see that 25/24 feet times 1.01376 equals 1.056 feet. This means that the equatorial measurement per day using 25/24 feet as a measurement of a degree will reveal the same number as the measurement in units of 1.056 feet between 51 to 52 degrees, the degree whose length is what every degree would measure on the mean Earth. In John Neal's metrology 1.01376 is called the standard geographical ratio. The measurement of a day on the equator will be 360,000 divided by 25/24 equaling 345,600, now in units of 25/24 feet. The degree at Stonehenge will therefore also be 345,600 units of 1.056 feet, and, crucially, the quarter degree between Stonehenge and Avebury will be 86,400 units of 1.056 feet (discussed in depth later in this chapter).

The whole of the future metrological system was therefore implicit in the AMY, a unit obtained only through day-inch counting over 19 years and resolved into 18/7 feet of 1.056 feet, the geographical foot commensurate with the degree of latitude that expresses the length of a degree on the mean Earth. This means that the use of the AMY, demonstrated in the inter-observatory distance from Lundy to Stonehenge, was significant. No other units or system of metrology were required to explain what was then made possible: From the measurement of the mean degree a calculation of the mean Earth circumference could be made, and from that, the length of the meridian of the Earth and of the radius of the mean Earth (taken to be 7/44 of the mean Earth circumference) could be found.

THE HENGE AS A MODEL OF THE EARTH

It has always been unclear why henges were built. Wikipedia tells us:

> The word *henge* refers to a particular type of earthwork of the Neolithic period, typically consisting of a roughly circular or oval-shaped bank with an internal ditch surrounding a central flat area of 20 m or more diameter. . . . The three largest stone circles in Britain, Avebury, the Great Circle at Stanton Drew stone circles, and the Ring of Brodgar, are each in a henge.[4]

Despite Stonehenge not being a true henge (its ditch runs outside its bank), it still includes all the necessary design elements to act as a circumpolar observatory. The banks of a henge can give a raised or consistent horizon and a deep inner ditch, as at Avebury, can allow low sightings across the inner flat area over which measurements and ropes could operate. In the case of Stonehenge, the ditch outside the bank enabled a raised sighting platform on which an observer could lie.

Stonehenge 1 was an *outer* ditch carved into chalk and a bank made of outfill. The circle of 56 Aubrey holes were cut into the chalk under grass, hence their longevity over 5,000 years. The monument was built at such a latitude that the northernmost moonset at maximum standstill of the moon was at right angles to the midsummer solstice sunrise. This fact was explicitly preserved by the later rectangle of four Station Stones that spanned the Aubrey hole circle with a diagonal of 13 units, each eight megalithic yards in length, to form the diagonal of a 5 by 12 unit rectangle. This rectangle perhaps became necessary when the center of the monument obstructed the sightlines originally made from the ditches through the center. However, as stated above, one of its diagonals is between special mounds signifying that the day-inch count between lunar maximum and minimum was important at the site.

To lay out the 56 hole circle of 104 megalithic yards, the circumference of a lesser circle, of diameter 100 AMY, could have been used to achieve division by the required 56 units, each of 365/2 inches and this would have been used to present a septenary model of the Earth's latitudes. The northern hemisphere would have 28 of the 56 holes, representing 14 latitudes, twice seven in each quadrant. One of these is the latitude of Avebury, built directly north of Stonehenge, at 4/7 of the quadrant, the eighth Aubrey hole north of east in the circle.* In terms of the whole planet that the circle represents, Avebury would be 360/7 degrees from the equator.

The geocentric latitude of Stonehenge had already been identified from the relative longitude of the Lundy to Stonehenge distance of 240,000 AMY, and we now know that ¼ of a degree separates the two. To measure the difference, the observatory must have been equipped with the ability to measure degrees, something not thought possible prior to the Sumerian culture who are thought to have innovated the degree circa 3000 BCE.

DISCOVERING DEGREES AT STONEHENGE

The easiest way to generate a circle with a perimeter of 360 units is to build a square of side length 81 units. As shown in figure 6.9 and in the metrological work of A. E. Berriman, the ratio between the side length of a square and an arc between two consecutive corners is 9/10.

Using the fine approximation detailed by Berriman, one simply scales up by nine times to obtain 81 on the chord and 90 on the arc. The diagonals of the square define a radius to generate the circle the square inscribes and this circle will be 360 units in circumference. Next, we can adapt the

---

*Within centuries of Avebury being located at latitude 4/7 north, the Egyptian city of Thebes was placed at latitude 2/7 (4th hole) and Delphi would come to be established at 3/7 (6th hole).

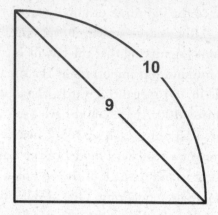

Figure 6.9. Using a quadrant of a square to create a circle of known circumference. The quadrant (chord/arc) length ratio is 9/10 if 22/7 is used for pi and 99/70 for the square root of two in ($2\sqrt{2}/\pi$). Both values were known in antiquity.

idea that the Aubrey Circle was intended to be 365.25 times 29.53 inches in circumference and realize that the units are day-inch counts for the lunar month, which are ¾ meter long. By doing so, a circle just within the Aubrey Circle emerges with perimeter of 360 lunar months or 30 lunar years, which equates to exactly 270 meters, as in figure 6.10.

The day count for a lunar month was probably handled as a type of megalithic yard (as explained in chapter 3's description of Gavrinis's stone L6) and hence divided into four parts to provide quarter degrees. On the circle of diameter 100 megalithic yards divided into 56 parts, useable to divide 360 degrees into 8 times 7 parts, this presentation of the quarter degrees would have allowed Stonehenge at 51° 10' 42" latitude to be seen as one quarter of a degree from Avebury's septenary latitude of 360/7 degrees (51° 25' 43"). This is a crucial point because only such geometrical tools, used alongside metrological knowledge, could have located the septenary latitude of Avebury as being one quarter of a degree from the latitude of Stonehenge, and when this was seen at Stonehenge, only then must the location for Avebury have been chosen so as to measure the quarter degree.

The megalithic were probably not able to determine an exact quarter degree of latitude between two sites and the quarter degree separation of Avebury from Stonehenge would not have been achievable without a Stonehenge capable of measuring its own latitude and seeing this latitude, in a circle of 360 degrees, alongside the septenary latitudes, symbolized by the Aubrey circle's fifty-six post holes. The distance between

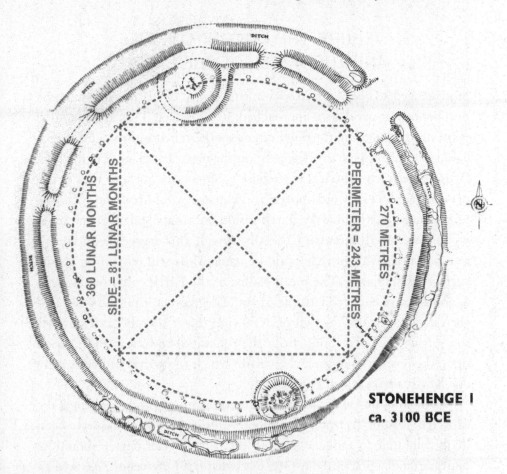

Labels within figure: DITCH, PERIMETER = 243 METRES, 270 METRES, 360 LUNAR MONTHS, SIDE = 81 LUNAR MONTHS, STONEHENGE I ca. 3100 BCE

Figure 6.10. How the Aubrey Circle at Stonehenge provided a means to measure degrees, when used as a model of the Earth's meridian given a circumferential length of almost exactly 360 lunar months in day-inches. Taken from the original survey plan, courtesy of A. S. Thom.

Stonehenge and Avebury, when seen on a 360 lunar month circumference at Stonehenge, would be a shockingly small 7.4 inches. However, using *scaling*, the whole circumference could be taken as the 90 degrees for the meridian quadrant so that Avebury would be at the 32nd hole of 56 and Stonehenge separated by a whole lunar month of 29.53 inches or ¾ meter. The measurement between them only had to be large enough to confirm *the existence* of a quarter degree of separation, between the next septenary latitude and Stonehenge.

AVEBURY, THE QUARTER DEGREE,
AND THE MODEL OF THE EARTH

Stonehenge had been located within that degree of latitude in which the number of feet per degree approached 365,000 feet. This 52nd parallel is the unique degree in the northern or southern hemisphere with length equal to the length of any degree upon the mean Earth, an Earth whose figure is perfectly spherical but whose volume is identical to that of the present Earth. The words north and equator would have no meaning on the mean Earth since the Earth would not rotate and all great circles would be perfectly circular. Crucially though, this degree gave access to a second measurement that could be made while still remaining on the surface of the Earth: The mean radius of the Earth could be deduced as the distance between latitude 51 to 52 degrees multiplied by 360 and then divided by 44/7 as two pi. Knowing this it then became possible to estimate the polar radius from the equatorial and mean radii, based upon the properties of the cube within which the northern hemisphere can be visualized as being inscribed.

In looking for the location one quarter degree from Stonehenge, the expected 365,000 feet of degree length on this parallel could have been divided by four to arrive at 91,250 feet. The quarter degree should be one quarter of a practical day (365 feet), that is, 91.25 feet or 33.6 AMY,* which can be applied exactly 1,000 times between Stonehenge and Avebury, locating the future site of Avebury at 4/7 degrees north. Such a use of thousands might have been the origin of using miles of 5,000 feet, which became a standard for the future metrological system. This is only possible *because* the astronomically derived AMY, based upon day-inch counting, divides into the mean Earth degree at Stonehenge.

Earlier we saw that the AMY can be seen in terms of a new foot, as 18/7 feet of 1.056 feet. This foot has a metrological mile of 5,000 feet that exactly matches the length of the English mile of 5,280 feet. The

---

*The actual figure of 33.604 AMY, if rounded to 33.6 AMY, gives the perfect result! The reason is that the difference between 33.604 and 33.6 is the same difference between 365,000 feet as the possible degree length and the actual degree length of 364,953.6 feet. This occurs because the AMY is commensurate with the actual degree length.

quarter degree then has 86,400 units of 1.056 feet, showing that this foot divides into the whole degree as 345,600 such feet. Dividing this by 5,000 gives 1,728/25 or 69.12 miles as the length of one degree of the mean Earth and of the actual degree between 51 and 52 degrees. Since 1,728 is 12^3 the mean degree is one-twenty-fifth of twelve cubed, a fact of significance concerning the later doctrine of cubic arks (see chapter 7).

The above would have estimated the mean Earth circumference to be 48,384,000 AMY, or 24,883.2 English miles of 5280 feet. This value is in harmony with historical geometry and close to the modern estimates, which can vary due to variations within the actual globe's ellipsoid. The number of feet in the mean Earth circumference (and the double meridian) is an amazing 44 times 12^6, which means that when pi is approximated as 22/7 and thus two pi as 44/7, this estimates a mean radius for the Earth of 7 times 12^6 or 3958.6$\underline{90}$ miles, which is 4.7 miles less than the actual equatorial radius.

Knowing these two radii allowed the polar radius to be found through the strange simplicity that emerges when one visualizes the box that encloses a three-dimensional quadrant (one-eighth) of the Earth, as in figure 6.11.

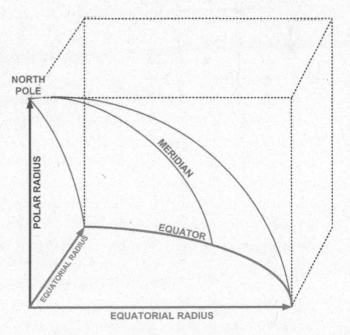

Figure 6.11. The cube enclosing a quadrant of the northern hemisphere has two identical dimensions along equatorial radii and the third is the polar radius. The properties of such a cube, common to all spinning planets, cause these radii to be systematically related to the mean Earth radius as follows: $R_{mean} = R_{pole} + 2d$ and $R_{equator} = R_{pole} + 3d$, where d is therefore $R_e - R_m$.

In this box we find two sides of equal length, both equal to the equatorial radius $R_{equator}$, and one side slightly shorter, equal to the polar radius R_{pole}. The volume is therefore $R_{equator}^2$ times R_{pole} and the mean Earth radius is the cube root of this volume, which is then:

$$R_{mean} = \sqrt[3]{(\mathbf{R_{equator}^2} \text{ times } \mathbf{R_{pole}})}$$

This equation always returns a mean Earth radius (R_{mean}) that differs from the polar radius (R_{pole}) by twice as much as it does from the equatorial radius ($R_{equator}$), a fact as true for Jupiter as it is for the Earth. This has the result that the three radii can be made to divide into each other using a common unit, exactly as one sees in the N:N+1 triangles used by the megalithic in their astronomy of time periods. Thus, if we take the equatorial and mean radii, deduced from the equatorial and mean Earth circumferences, their difference in feet would have indicated a common factor between them. This common factor could then have been doubled to obtain an estimate for the size of the polar radius. These three radii are related as

$R_{mean} = R_{polar} + 2d$ *and* $R_{equator} = R_{polar} + 3d$, *where* $d = R_{equator} - R_{mean}$

The model will then be in the form N:N+2:N+3, where N equals the polar radius, presentable in a right-angled triangle with an intermediate hypotenuse.

If the megalithic had visualized this, they would have seen that to make use of this relationship required two of these radii to be found in order to establish the third. This preemptive thinking about the organization of their geodetic challenges reveals their megalithic genius at work in solving apparently impossible problems. The final part of their genius lies in their use of the astronomical megalithic yard as 18/7 units of 1.056 feet, revealed in the length of the quarter degree between Stonehenge and Avebury, which contains exactly 86,400 of these 18/7 units. The difference between the equatorial radius and the mean radius, when divided into each of them gives a whole number relationship based near N = 840. The amazing fact about this is that a suitable factor, 864, was available as one-hundredth of the quarter degree. This

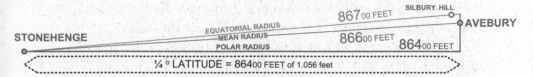

Figure 6.12. The geodetic triangles formed between Stonehenge and latitude 360/7 degrees 86,400 feet (of 1.056 feet) north of Stonehenge, Avebury (henge), and Silbury Hill, which is 86,400 English feet from Stonehenge.

was then usable as the factor within a model of the Earth where 100 units of 1.056 feet (105.6 feet) could be equated with the difference between the equatorial and mean radii of the Earth, to model that difference within a geometrical arrangement.

To model the three radii, the 86,400 "new feet" between Stonehenge and the Avebury ridgeway directly to the north was made to represent the smallest Earth radius, the polar radius. This meant that a line 86,600 feet from Stonehenge, representing the mean radius, could be angled to form a right triangle with the northern length representing the polar radius. A further line of 86,700 feet could then represent the equatorial radius of the Earth.

The line representing mean Earth radius terminates within the henge of Avebury and the equatorial line terminates further west, in the garden of the present "manor house" after traveling through Silbury Hill, the site of which was carefully placed at 86,400 English feet from Stonehenge.

PI IN THE EARTH

At this point one can note a conjunction taking place between circular built structures studying a slightly non-spherical Earth and units of measure that have developed the relationship between a radius or diameter and its circular perimeter. This conjunction leads up to the language of pi, which, for all intents and purposes, is the informing god of transformations between the rotational world of the sky and the linear dimensions found upon the Earth. Linear measures then developed into a science of intentional measure, namely metrology. Something

about pi naturally informed megalithic geometers about the characteristics of rational fractions that could approximate this ratio.

In my own journey I have come across many integer fractions for pi
that surprised me, such as 820/261, apparently understood at Carnac and
able to convert diameters of megalithic yards (261/8 inches) into perimeters that divide by 82 inches. But such values of pi stand disconnected by
the large prime numbers involved as factors and so, as the numerical sen-
sibilities of the megalithic developed, attention turned to using no prime
numbers higher than 11. In other words, the five lowest primes, 2, 3, 5, 7,
and 11, and the most accurate and simple value of pi, 22/7, were evidently
used and relatable to septenary units such as 12/7 feet, the royal cubit,
and 18/7 units of 1.056 feet in the AMY. The latter foot of the AMY has
the rational formula 132/125 feet, which then places 11 in the numerator
and removes seven in the denominator, relative to the English foot.

One can see from these early steps how the English foot would
come to dominate a metrology of rational fractions and how the prime
numbers within 132/125 (especially 11) are somehow dividing into the
degree length from Stonehenge to Avebury. This is seen in how a 5,000
foot mile of 1.056 feet equals the 5,280 foot mile of English feet, yet it
is hard for the modern mind to see what is happening because it really
should not be happening within out modern view of causality. The key
in the case of the English mile is that it adds 11 to a measurement,
while 11 can also be added within the fractional feet used by metrology. Therefore, things were being discovered about the Earth itself by
using fractions with only small prime numbers as their factors, and this
enabled latitudinal degrees, circular structures, and the dimensions of
the Earth to all become rational in ways that involved pi.

What turned 21/20 feet into a foot that divided the mean degree
at 51 degrees was the ratio 176/175, implicit within the AMY, itself
naturally generated by measuring the excess of the solar over the lunar
year over 19 years using day-inches. This ratio of 176/175 is a composite
of two versions of pi since 176/175 equals 4/25 times 44/7. John Neal
suggests that ancient metrologists would have originally been drawn to
this ratio, as it gave mastery over the symmetrical worlds of the number
within a measurement versus the length of a unit of measure in both
the radius and the circumference of a circle.[5]

The Operation of Pi within 176/175

The less accurate pi within 176/175 offers a simple way to compare a circle's radius and its perimeter, but the inverted two pi of 4/25 is therefore being projected backward, from the perimeter back to the radius. The accurate two pi of 44/7 was therefore thought to act from radius to perimeter and was being used as *the actual ratio of pi*, while the inaccurate ratio of 25/4 was offered as a convenient, though deficient, ratio for pi. The convenience comes in that one can declare any number one wishes on the circumference, then divide this by 25 and multiply by 4 to establish the required radius, measured in units larger by one-175th part. This radius will then describe the chosen number of whole units on the perimeter, maintaining rationality through using two related units of length.

The utility of 176/175 could have been realized quite naturally from the AMY, which is itself 18/7 units of 1.056 feet: If counted in such feet one would discover that 175 units of 1.056 feet equal 176 units of 1.05 feet. These two numbers could themselves be easily factorized, 175 being seen to contain seven to leave 25 and 176 to contain 11 leaving 16. Halving 16 twice reveals the familiar pi values of 25/4 and 7/44, as the presence of two different (and hence nearly cancelling) pi values. Self-cancellation revealed a subtle magic in which the longer measure used in the radius (measured in units larger by one-175th part) could lead to a number of units made up of a whole number of quarter units of the shorter variety. The English foot and all other feet came to be varied in this way within historical metrology.

Another anomaly involving pi that made an enormous difference was found in the unlikely guise of performing factorizations in the form of a cube in order to evolve the notion of volume. We have seen how visualizing the AMY as 18/7 of a hitherto unknown foot was achievable by seeing a royal foot as cubit that is again made half as big (multiplying by 3/2 as in figure 6.7 on p. 156). This means the unit represents a volume, at least conceptually, and since managing the numbers within measured lengths required a firm grip on the factors to be found within each number and within each version of pi, then a general interest in volumes would not have been overly theoretical.

A problem of great interest arises when ideal cubes, of three equal side lengths, are needed that double in volume by an equal increase in their unit side length. This was an ancient Delian challenge[6] credited indirectly to Apollo, whose altar, on the Island of his birth, Delos, needed to be doubled in volume. The answer today is to get a scientific calculator to find the cube root of two, so as to see that the answer requires a side length increase of 1.26 or from a side length of 100 parts to one of 126 parts (to an accuracy of one part in 16,000). The doubling of a cube's volume can then be achieved by increasing the side length by 1/5 but measured out using a variation of any foot that is 1/125 larger (1.008) than the measurement of the original side length. Almost all the units of length known to historical metrology had slightly varied versions exactly 1.008 times different in length.

This ratio of 126/125 is again related to two approximations to pi:

$$63/10 \times 4/25 = (18 \times 7)/5^3 = 126/125$$

And so, the two ratios 176/175 and 126/125 appear to have been made available within metrology because they again had uses for circles that have rational radial and perimeter lengths as well as for doubling the volume of a cube. The difference between these two ratios was also used, namely 441/440, which of necessity cancels the good pi of 44/7 and the bad pi of 63/10, thus forming a square of variation within units of length. Figure 6.13 illustrates John Neal's discovery that any rational fraction of the English foot could form a root measure and important variations of both root and standard measure using these two fractional relations. Neal's terminology for these varied measures is retained: "canonical" for a single variation of 176/175 and "standard" for a single variation of 441/440, leading to a combined "standard canonical" when a measure has been altered by 126/125.[7]

After this preview of how metrology would come to develop, one can see the foot of 1.056 feet, derived from the AMY as 18/7 of those

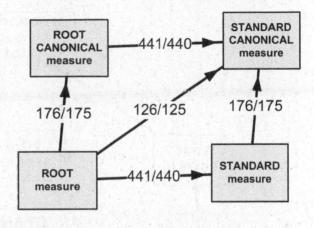

Figure 6.13. How rational fractions of an English foot form a root foot that could be varied by two ratios, leading to a square of slightly different measures, meaningful for building circular and cubic structures and expressing different approximations of pi.

feet, was later termed the standard canonical Persian foot, since in historical times the root foot of 21/20 (1.05) feet was named after its discovery in lands of the Arabian empire's Persian dominions.[8] However, the magic ability of this microvariation of the Persian foot to divide into the average degree of the mean Earth (and hence the meridian) does not come from the Persian foot at all but, in my view, from what appears to be the older root of that foot, 25/24 feet, which could be called the inverse Roman foot in the absence of any historical record for its existence.* I would term it the "original varied foot," since it is operational around 3000 BCE as perhaps the first foot that would create a geographical foot, vital to this story and to history.

Within the module of a given root foot, a geographical version is defined by multiplying a standard canonical foot by 126/125 (1.008) and then varying it again by one-175th part to create a standard geographical foot, the best known example being the geographical Greek foot of 1.01376 feet. This has varied the English foot as its root of one foot and hence exposed the geographical fractional ratio in all cases of the root times 3,168/3,125 (1.03476). The standard canonical variation of this is the product of two versions of pi, 63/10 times 4/25 (1.008†). Now a further multiplication by a different pair of pi approximations,

---

*Neal would call 25/24 (1.0416) feet a "Lesser Reciprocal Persian," but I am talking of a time when the later metrology was still taking far simpler steps.

†The foot of 1.008 feet is known to historical metrology as the Olympic foot, found widely in the Classical world.

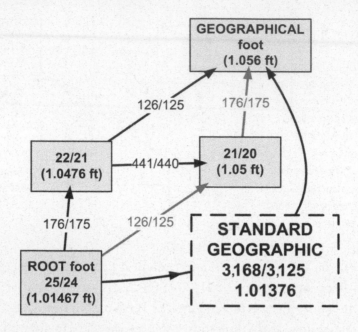

Figure 6.14. How a geographical foot, 1.056 feet, emerges from the unique root foot of 25/24 feet, belonging to the only three rational feet, which are related by the three metrological ratios of 176/175, 441/440, and 126/125.

4/25 times 44/7 (176/175), leads to the pi of 25/4 having been squared and seven having been removed altogether, leaving just 11 above, in 3,168. The formula is $3,168/3,125 = (4/25)^2$ times 63/10 times 44/7, a sort of quadratic of different rational approximations to pi. As a rational fraction, 3,168/3,125 is the ratio between the number of feet at 51–52 degrees and the number of feet in an angular day on the equator. In the next chapter we will see how John Michell identified the mystique of the number 3,168 as being associated with the anciently defined perimeter of any temple, making any temple perimeter associated with the north-south meridian and hence to the mean Earth, a kind of spiritual Earth in which temples aspire to place people.

This geographical ratio had turned up as the foot of 1.056 feet, courtesy of the sun, moon, and Metonic period generating the AMY, which is 18/7 of it. When measured using day-inch counting, its root foot came to be known as 7/18 of the AMY and was called "geographic" because it would then divide into the quarter degree between Stone-henge and Avebury. But its rational root foot was 25/24 English feet, the only ratio that with two other rational feet, expresses the necessary geographical ratios:

176/175, which equals 24/25 times 22/21, and
441/440, which equals 21/20 times 21/22, leaving
126/125 as 24/25 times 21/20,* which cancels the root of 25/24
  feet to then equal 1 times 21/20, the foot of 1.05 feet.

The metrological ratios are therefore based upon three different rational approximations in the value of pi, and these are uniquely expressed between only those three ratios: 25/24, 22/21, and 21/20, within the number field itself. The ratio 22/21 is itself one third of pi as 22/7, and the further application of 176/175 to 21/20 gives the geographical ratio relative to root 25/24 feet, in which all three values of pi are employed to give a ratio from the double cancellation of pi.

It appears, therefore, that all of future metrology emerged from the AMY, which was used in Britain to locate Stonehenge. The AMY itself had emerged from day-inch counting over the 19-year Metonic period. The next chapter will show how that system of metrology indicates that the Earth itself came to be shaped according to this pattern of approximations to pi and root feet, discovered originally through a geometric analysis of the astronomical megalithic yard.

---

*See my *Sacred Number and the Origins of Civilization*, page 42.

7

FROM EGYPT
TO JESUS

For what shall it profit a man
if he shall gain the whole world
and lose his own soul?

THIS QUESTION, ATTRIBUTED to Jesus, could ironically relate to a megalithic culture that had developed a model of the whole world but somehow lost cultural purchase after the invention of writing, great empires, world religions, and, these days, industrial globalizing technologies.

Egypt was strongly founded upon the model of the Earth and, over the course of two millennia, their pharaohs ruled as god kings. Outside their domain the sacred power of kings was in competition with the growth of middle-class guilds of specialists—that is, until Alexander the Great (356–323 BCE), the last ancient god king, created an empire that would morph into the Roman Empire.

This divinity of the king obviously emerged due to religious ideas

Figure 7.1. (Opposite) Photo of *The Wailing Wall* by Gustav Bauernfeind (1848–1904). The Wall is all that remains of Solomon's First Temple, and it was extended by Herod the Great to provide a perimeter wall around the Temple Mount one Greek mile in length, this being one-25,920th of the meridian of the Earth.

and these could have been a transposition of megalithic knowledge, of astronomy and geodetics, into religious ideas about rulership and destiny. Human individuation was generally subsumed within a god king's splendor, but by Greek Classical times individualism was seen emerging alongside numerous cults and mysteries. But the Roman emperor Constantine (306–377 CE) would eventually either integrate these cults within a Roman Christianity or suppress them.

Jesus, an elusive but definite historical figure, was said to have been of royal descent, of the line of David. Jewish kings did not take center stage, but rather the history of the Jewish people's covenant with God was written in Babylon as a retrospective called the Bible. In 957 BCE Nebuchadnezzar, the king of the Neo-Babylonian empire, sacked Jerusalem and destroyed the First Temple built by King Solomon in 586 BCE. The king then exiled the Jews over the next decade to his capital. At this time, when their history must have appeared to have ended, scribes wrote the Bible, which, through its artifice, combined memorable stories with state-of-the-art techniques in literary composition and this firmly established the identity of the Jewish people within history. Their retrospective work, including patriarchs, judges, kings, and prophets, held nothing to be greater than their single god, YHWH.

In contrast, the history of the Egyptians was the story of a state ruled by pharaohs who were manifesting a divine aspect of time called Horus. Horus could enter a royal individual and manifest history as the *ipso facto* destiny of their people. At death the king's mouth was opened to let out the spirit (Ka) of Horus while the body was embalmed and placed in a chamber for life in Eternity.

Unlike the spiritual beliefs of the Egyptians related to the divinity of the king, the Bible detailed a covenant (a contract) in which God was making history through the Jewish people, and so, when Jesus spoke around 30–33 CE the influences upon this people of the covenant had been:

a) All were descendants of a common patriarch called Abraham of Ur of the "Chaldees" (but probably a similarly named Akkadian outpost within the Hittite empire).

b) The idea of YHWH became clearer through three generations of nomadic life in Canaan, involving various experiences and teaching stories.

c) Twelve children were born of third-generation Jacob, renamed Israel, one being Joseph who, through a fateful betrayal by his brothers, becomes a royal advisor allowing the children of Israel to settle in Egypt.

d) Moses is born in Egypt, in the standard motif of being a future threat, placed in a floating basket (an Ark), only to be found and adopted by a royal princess.

e) After being educated in the craft of a royal priest, Moses leads his now oppressed people out of Egypt and "wanders" for decades through the Sinai desert, then communing with YHWH on Mt. Sinai and establishing religious laws and the dimensions for the sacred tabernacle.

f) Canaan is conquered under Joshua, when the Phoenician alphabet might have first been adapted to become Hebrew as early as 1000 BCE.

g) There were periods of Judges, Kings, and Prophets. YHWH instructed King David to build a sanctuary on the Temple Mount to replace the Tabernacle and serve as Temple of the Israelites.

h) Judea was defeated by Nebuchadnezzar II, the temple destroyed, and the Jews made to live in Babylon, where the Bible was composed, asserting the historical reality of the Jews.

i) There was a return to Jerusalem when the Babylonian king changed.

After all this but preceding the time of Jesus, over one hundred years of messianic fervor arose along with some new prophets. The Roman client-king of Judea, Herod the Great (74–4 BCE), started building a Second Temple around 20–19 BCE.

The builders of this Second Temple were most likely connected to a geodetic tradition that escaped from Egypt, probably through a figure like Moses. Thereafter, metrology became an important component of the Jewish religion, of statecraft, and of the art of sacred building. In other words, there were megalithic and Egyptian precedents to Jewish temple building and, it appears, to this Second Temple into which Jesus then fits perfectly as a historical figure. The written history of the Jews made them a unique culture among the other ancient cultures of the Near East.

Many Jewish stories were borrowed from the Sumerians and Babylonians, one such being the story of Gilgamesh who met Utnapishtim, the first builder of an ark with definite dimensions. The original Temple of Solomon stood on Mount Moriah to the east of the old city, itself a dimensional construct to hold the Ark of the Covenant that had been with Moses in Sinai.

The Bible says the Ark and original Tabernacle around it were defined by YHWH in cubits, that is, in units of three halves of a foot. This foot was almost certainly an Egyptian royal foot of 8/7 transformed from its cubit of 12/7 feet by 126/125 into the cubit of 1.728 feet, the cubit found within the design of old Jerusalem. The story of Moses and the presence of this measure in Jerusalem point to a system of Jewish measures derived from Egyptian measures. Ancient sacred buildings relied on these sacred measures exactly because of their geodetic connotations. For example, the Jewish sacred rod was one–ten millionth of the Earth's polar radius. The mystique, if not power, of sacred measures came from their being related to the Earth, in units rational to the model of the Earth determined in megalithic times.

YHWH AS A GOD OF HISTORY

From the above one can deduce that the idea of sacredness, of kingship, and of the manifestation of god as history were conflated as belonging together in the ancient world. The apparent catastrophe of the destruction of the First Temple and the subsequent exile of the Jews appears to have triggered an unprecendented creative task, to write the Bible as their crystallization of their ancient world. As Cyrus H. Gordon notes, "YHWH is a God of History like no other," creating a unique historical cult:

> This creation of linear history is an innovation of ancient Judaism and it is directly related to its unique understanding of God. Since

YHWH is a God of History, interacting in human affairs, the telling of history becomes a sacred act. Among the other peoples of the ancient world, whose gods were gods of Nature, the creation of historiography or of a document such as the Genesis-Kings account was non-existent. Israel's neighbors had all the features of this large collection (creation stories, flood stories, epics, law collections, tales of wonder, royal annals, etc.), but it took Israel to put it all together into one unified work.[1]

According to the biblical account, the god of the Jews revealed his will through various types of people as well as through angels and dreams, happily reflecting Persian ideas within the stories, none more so than the messianists who preceded Jesus and would have seen the rebuilding of the Temple Mount as confirmation of prophesies concerning a regeneration of the covenant through a coming messiah.

The first phase of building was a refurbishment of the Temple Mount, as suitable for a Second Temple. This involved a doubling of the area of its single hilltop terrace by extending the terrace to the south.* Parts of the western boundary wall from the First Temple are still standing (now the Wailing Wall). A new southern wall was built by Herod, whose architect was probably a member of the former dynasty and preserved geodetic knowledge in the perimeter length of this walled enclosure. So, while great attention has been given to the exact design of the Second Temple and its dimensions, the perimeter wall has been taken as just a necessary enclosure and support. But the new walls around the Temple Mount present a perfect model of pi, of the meridian length, and of the mean Earth radius, while also dividing the meridian into the number of years anciently assumed for the equinoxes to complete their precessional cycle around the zodiac.

It was English antiquarian John Michell who noticed that the old western wall had, in its extended length, come to represent the diameter of the mean Earth.[2] All four sides of the new wall taken together and then multiplied by 12,960 would equal half the circumference of the

---

*Under this platform was built a set of arched columns that became known as the Stables of Solomon, in which the Knights Templar famously lodged.

Figure 7.2. The new wall built by Herod as it is today. The western wall is to the left and leads to the Wailing Wall. The south wall, here illuminated by the sun, allowed the platform of the Temple Mount to be extended.

mean Earth, which is the length of the actual north-south meridian of the Earth, despite the slight deformation of the spherical mean Earth due to rotation.

The perimeter of the new Temple Mount enclosure was 5,068.8 feet, which is the length of one Greek geographical mile, a measure found throughout the Classical world. The scaling factor of 12,960 to the meridian is half the number of years in the ancient value of 25,920 years, the now familiar duration of the Precession of the Equinoxes, a value firmly established in the ancient world. The Jerusalem cubit was 1.728 feet long and so the Jerusalem foot, taken to equal one unit compared to the cubit of 3/2 units, is 1.152 feet long.*

Through this John Michell found that the new western wall of 1,400 such feet added to 880 (northern wall), 1,320 (eastern wall), and 800 (southern wall) gave 4,400 Jerusalem feet as the perimeter, so that the ratio of perimeter to western wall, 4,400/1,400, gave the primary rational approximation of pi, 22/7 (see figure 7.3 on p. 180). Also, one sees that 4,400 times 1.152 is 5,068.8 feet, exactly one Greek mile. Therefore, if the western wall represents the diameter of the mean

---

*This unit of 1.152 is actually the foot of the Egyptian royal cubit of 12/7 feet transformed by the standard canonical ratio of 126/125 (1.008).

Earth and the perimeter of the Temple Mount enclosure is one Greek mile, we can scale back up with the scaling factor of 12,960 and find that the full circumference of the mean Earth is exactly 25,920 Greek miles long—an observation also made by Michell.[3] The metrology of the Temple Mount's perimeter therefore connects the wall to the emergence of precessional history at the exact period when the world was to move from the age of Aries to the age of Pisces and when any Messiah might be conflated with the new Lord of the Age, already established within a number of mystery cults.

We have now exposed an important clue as to how Great Time came to be related to the ancient model of the Earth. Though all physical meridians spread out from pole to pole as lines of longitude, two key meridians were different. Visualized as halves of a great circle and connected to the equator at the equinoxes, these two meridians were not on Earth at all but rather in the sky, in what the ancients called the celestial Earth. These two equinoctial meridians were seen as forming a great circle that passed through the two equinoctial points of spring and autumn at right angles to the celestial equator, unlike the ecliptic, which also passed through the equinoctial points but at the angle of the Earth's tilt relative to the solar system.

This simple visual model of the celestial Earth was based upon

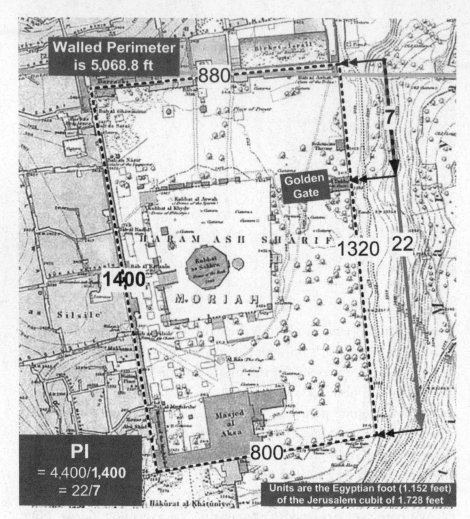

Figure 7.3. The perimeter wall of Jerusalem's Temple Mount showing a pi ratio of 22/7 between western wall and total perimeter and between eastern wall and Golden Gate. The perimeter is 5,068.8 feet, by then the Greek geographical mile, and one-25,920th of the mean Earth circumference. Adapted from John Michell, *The Temple at Jerusalem.*

time and involved just two such rings (shown in figure 7.4), the ring of the celestial equator and, passing through the poles, the ring formed by the two equinoctial meridians. Both were great circles that were divided metrologically so as to represent time: the equator into 365.242 units of 72 miles (each of 5,000 English feet) and the two meridians together into 360 units of 72 miles (each of 5,068.8 feet).

Since the equinoctial points move backward through the zodiac, the meridian ring would be seen to move westward each year within this celestial model. On the equator this would amount to about one mile of 5,072.8 feet per year. At the same time the meridian was seen to drop at the spring equinox by one Greek mile of 5,068.8 feet and therefore rise by the same amount at the autumn equinox. That is, each year there are new spring and autumn equinoctal points on the celestial equator, spring descending and autumn ascending, as also stars above the celestial equator will drop below that equator and others rise up from below.

This celestial model with movements related to the equinoxes was

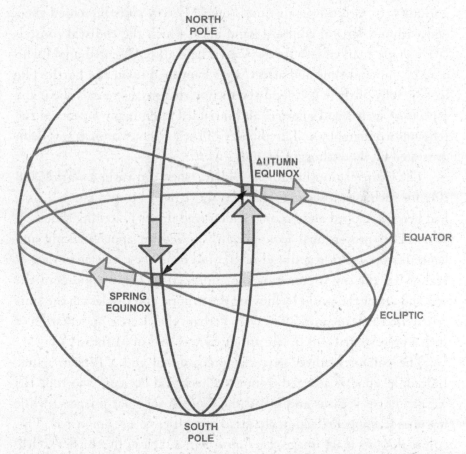

Figure 7.4. The conceptual model of Great Time as two bands at right angles, joined at the equinoctial points where the celestial meridian of the Age crosses the equator and ecliptic.

like a precessional clock. However, despite its metrological connections to precessional time, the Temple Mount could not reflect these same movements of the precessional clock. But, being an established center of the world (or *omphalos*), Jerusalem stood forever on the meridian of the Age, though rotating once per day. Its degree had the radius of curvature belonging to the mean Earth, the spiritual antecedent of the meridian.

THE MANY FACES OF PRECESSION

This model of precession, a north-south Ring of Ages, explains an enduring feature within at least some of the Mysteries found in late antiquity. In Greco-Roman times some Mystery cults presented their god within a vertical elliptical band upon which the zodiacal constellations were marked (see figure 7.5).[4] These all represented a god who could command time and affect what happened upon the Earth. The iconography of these gods and their associated myths reveals the many aspects of such a god's powers and intruded upon many aspects of the religious iconography and mythology of the ancient world, as was demonstrated by the authors of *Hamlet's Mill*.

The clear relationship to the zodiac, when shown in a vertical setting for such a god, suggests that the above model of precessional time had been developed and become widespread in the preceding thousand years. This precessional iconography showed an epoch-making and hence history-making godhead who could bend the everyday powers of Helios-Ra, the sun, by moving the equinoctial crossing points of the sun and affect the world by moving the ecliptic and displacing the stars relative to the framework of heaven. Perhaps the clearest presentation of such a precessional cult in late antiquity was that of Mithras.[5]

The cult of Mithras gave the precessional god a Persian name. Indeed, he appears adapted from the Greek god Perseus, who rode the winged horse Pegasus and killed the Gorgon (of solar imagery) while rescuing Cassiopeia (held in chains as a sacrifice to a sea monster). This cult took as its main image the *Tauroctony*, a tableau in which a youthful Mithras, in his signature Phrygian cap and cape, is killing a bull while looking away.

Figure 7.5. The widespread tradition of a God who changes the astrological Age, through the Precession of the Equinoxes: Top left, Mithras as Sol Invictus; top right, Mithras slaying the Age of Taurus; bottom left, Aion, God of Ages; and bottom right, Orphic God Phanes. Mithras slaying the Age of Taurus photo by Tim Prevett courtesy of the Segontium Museum, 2005.

In the tableau Mithras has two important attendants to right and left. Both have their legs crossed indicating the points of crossing for the sun on the celestial equator and each carrying a torch that represents the movement of the sun at each of these equinoxes (see figure 7.6). On the right side the figure, Cautes, represents the spring

Figure 7.6. Tauroctony in Kunsthistorisches Museum, Vienna. Precessional god Mithras presents his killing of the bull Taurus, at what was then the spring equinox. He is shown between his two equinoctial Dioscuroi, spring on the viewer's right and autumn on the left, Cautes and Cautopates. The diametrically opposite constellation of Scorpio is shown below.

equinox, with the sun torch pointing upward to show that the summer lies ahead. The light of Cautes' torch is shown touching the mouth of the bull of Taurus, indicating the scene takes place at the spring equinox. On the left side we find Cautopates, the fall equinox, whose torch is pointing downward to the left indicating the winter to come. At the feet of Mithras is the Scorpion (the constellation of Scorpio) and serpent carrier Orpheuchus, both diametrically opposite the constellation of Perseus in the sky.

In the constellation of Perseus, we find Perseus holding the Medusa's Head (the variable star Algol), which is iconic of the sun in Taurus at the

Figure 7.7. The constellations of Perseus (top) and Serpens/Orphiuchus (bottom), opposite each other in the celestial Earth. They formed part of the new meridian after the vernal equinox, and hence the precessional meridian, left Taurus. From Urania's Mirror, a boxed set of 32 constellation cards engraved by Sidney Hall, published in 1825.

spring equinox being cut off by the precessional god who stands above Aries, the new world Age. Perseus is therefore part of the vertical *colure* or meridian of that age, presented upon the celestial Earth of the stars. Such stellar iconography developed without sculpture or scriptural myths, since the night sky and its named constellations could illustrate the story.

The well-educated Jesus stood at the end of a dying age. Within the walls of a new Temple Mount, he was aware that the Jewish people were expecting a new dispensation and also that the Mystery sects, popular with the Greeks, Romans, and Jews, were actually religious remnants of an astronomical fact, that the equinoctial sun had moved on before and was moving now from Aries to Pisces. Within four centuries the Roman empire would be declared Christian, but this would involve a Christianity quite unlike the early Church, a Jesus quite unlike the original person, and a Christian scripture quite unlike those that circulated among early Christians. Where Mithras, Aeon, Phanes, and Sol Invictus had once stood, Jesus would come to stand, as "Christ in Majesty," shown with the same precessional iconography that once accompanied the earlier precessional deities.

Figure 7.8. The central western gate at Chartres Cathedral holds Jesus as Cosmocrator exactly following the form of the ancient Mystery cult precessional iconography. The elliptical has been replaced by a vesica pisces, especially suited to the Gothic style of this early thirteenth-century cathedral.

On early icons Jesus was depicted as Lord of the World, the Pantocrator or Cosmocrator (see figure 7.8). He was shown on a majestic throne, usually with his feet on an orb representing the Earth, within a vesica pisces (instead of the elliptical of Mystery cult iconography) and surrounded by the Evangelists, these being openly associated with the fixed signs of the zodiac: Aquarius (the man), Leo (the Lion), Taurus (the Ox), and Scorpio as an Eagle.* Jesus came to be known in the early Church as the FISH because the Greek for fish ΙΧΘΥΣ (Ichthys) is an acronym for Ιησοῦς Χριστός, Θεοῦ Υἱός, Σωτήρ, (Iēsous Christos, Theou Yios, Sōtēr), which translates into English as "Jesus Christ, God's Son, Savior." This iconic acronym was compatible with his being the Lord of the Age of Pisces and a fisher of men. His mother was represented, within the first three gospels, as the constellation opposite to Pisces, Virgo the Virgin. "The Virgin" was a common motif within Greek myths of parthenogenesis (παρθένος, *parthenos*, meaning "virgin," and γένεσις, *genesis*, meaning "birth"), applied when a god gives birth to a hero through a mortal woman or as when Athena emerged from the head of Zeus.

There needed to be sufficient cosmological compatibility for an agreed form of Christianity to be accepted by the adherents of the other important Mystery schools as the single religion of Rome. Thus Jesus came to be born on Mithras's birthday of December 25th, three days after the winter solstice. This must have reassured the initiates of the Mithraic mysteries, many of whom were Roman legionnaires or other people of position in the Roman Empire. Many other signature features of Mithras were incorporated into Jesus as well, such as the virgin birth, his death at the age of thirty-three, and so on, so as to make him recognizable as a precessional hero, something not widely recognized today.

FRAMEWORK CONDITIONS FOR PLANET EARTH

We can now ask what it is that connects the Tai Plaque of the Upper Paleolithic with the symbol of the Cosmocrator around the beginning

---

*An Eagle because Scorpio's meridian includes Ophiuchus the serpent carrier and Vega the descending eagle.

Figure 7.9. Christ in Majesty painted in 1408 for the Cathedral of Dormition in Vladimir, Russia, by Andrei Rublev and Daniil Cherny.

of the current era. On the level of their materiality or function, they are both tools along the path of human cultural evolution, a human world that must be discriminated from the tools of Nature (such as bacteria and insects) and also from the tools of the Creation (such as stars and planets), the latter having made our living world.

Living beings are Nature's tools, part of its technosphere for building ecospheres and biomes. Our planet and its moon are solar system tools, a planetary technosphere for building a planet suitable for life. Planets evidently need tuning up to establish life properly, requiring a moon, special ratios in size, orbit, rotation, and, in the case of the earth, a tilt and consequent precessional periodicity.

Human tools arise within Nature to create something new, and these belong to a human technosphere, consisting of tools invented by man. Without an environment of human tools, nothing of any cultural interest, different from Nature, would arise within the human world.

Like language, writing, recorded knowledge, measurement, arithmetic, and decoration, tools enable meaning to be established and recorded. This world of meaning-making differs, though, from the functional world of tools because it involves the human imagination, cultural memory, ethics, and aesthetics. Meaning-making is founded on the Word or Logos and can therefore be called a *logosphere,* which defines the extent of all current meaning-making, including the meaning given to the past.

The Tai Plaque's impact upon the modest logosphere of the Stone Age was probably minimal though significant, as it indicated a tradition of time-factored bones and at least latent numeracy. This has been amplified in our time by Marshack's analysis of its meaning so that this plaque can be seen as a crucial precursor to the astronomy and geometry of megalithic in northwestern Europe.

I believe a different kind of significance can be found within the iconography of the Christ in Majesty cosmocrator image. The design arose within a well-developed logosphere (the ancient Near East), in which many signs and symbols were drawn together from 3,000 years of texts, religious influences, and iconography. This image was usually carried on a small piece of planking so as to be portable like the Tai Plaque and it was intended to connect its owner to the sacred. As we have seen

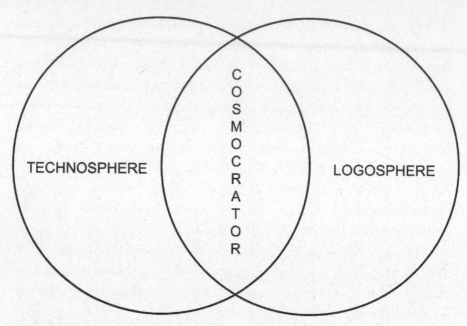

Figure 7.10. The circles of technosphere and logosphere interpenetrate to form a synthesis of functional cosmic clockwork and meaningful patterns, the cosmocrator, represented through the sacred geometry of the vesica pisces.

earlier, the ancients had an exact and highly developed sensibility of art with a conscious purpose. Let us see what might still be received from our cosmocrator.

The two "circles" of technosphere and logosphere interpenetrate to form a common presence, symbolically represented through the sacred geometry of two overlapping circles. This geometry connects the viewer to the cosmocrator who, as a synthesis of both functional cosmic clockwork and a logos of meaningful patterns, inhabits the central shared portion of the circles, the vesica pisces. Images like the icon of *Christ in Majesty* are a wager on transcendence* in that they connect to a world of human potentiality, a potentiality given to this planet as a creation in which free will *can* arise. To create the right conditions for free will requires there be a *cosmic* technosphere, a planetary tool for enabling the arising of a will independent from that which created the planet.

*According to George Steiner, "a wager for transcendence" is "when we encounter the *other* in its condition of freedom."[6]

The cosmocrator represents this cosmic technosphere as the will within creation for free will to arise, a Savior.

The logosphere that formed over the course of three millennia in the ancient Near East can be extrapolated as having interpenetrated a cosmic logosphere, made up of meanings held within the divine world itself. The development of religious meanings from astronomy and the geometry of our planet forged a set of fundamental religious ideas rooted in the structure of the cosmos. The sacred image of cosmocrator implies the sacrifice of the Lamb of God, *Agnus Dei,* but shows his subsequent resurrection as the Lord of the Age, with the necessary power to shift the cosmic clockwork. A constant renewal of the cosmic backdrop is redemptive for human existence because it prevents the unchanging dominance of any single cultural influence, somehow connected with the stars, and so renews the potential on Earth for free will through changing circumstances.

Though a product of a human imagination, our world of developed meaning is grounded in the structure of the world and our striving to be free. Religious meaning emerges from suitable acts of imagination in which the self (potentially free) may emerge through the otherness of (what can only be imagined as) a spiritual world, a world designed to bring about such freedom. It is for this reason that a human projection of our own potential free will takes the form of the God who created a world in which solace might be found. Such acts of imagination may seem arbitrary or unconnected to anything definite if significance is only given to our technospheric world as if it were a complete picture of reality. The results of free will are only registered after they emerge within existence, for they are *acausal,* or beyond ordinary time, and hence thought miraculous to those who can see them at all.

It therefore seems reasonable to propose that megalithic peoples retrieved aspects of the cosmic technosphere while the peoples of the ancient Near East developed pictures of a cosmic logosphere, during a period of religious experimentation. The region where these cultures overlap is found in the precise precessional metrology of the Temple Mount walls and the meaningful symbolism of the cosmocrator depicted in iconography contemporary to Jesus, a synthesis impossible until both the megalithic technosphere and the Iron Age Near East logosphere had been sufficiently developed.

THE ARK OF THE COVENANT

One of humanity's oldest and most enduring symbols is of an ark into which things of value are placed in order to save them from an impending flood, a shared and quite universal theme found in many different mythic traditions. An ark escaping from an antediluvian period, "before the deluge," is most widely known in the Bible story of Noah's Ark, which was based upon an earlier story included in the Sumerian *Epic of Gilgamesh,* a tale about the fifth king of the Sumerian King List, who reigned for 126 years (a key number for doubling the volume of a cube and present in measures varied by 126/125).

Like the seven days of creation, this first ark was copied by the writers of the Bible from Babylonian cosmology, which was inherited from the Sumerians. The written language of cuneiform, which emerged from the Sumerian invention of writing on clay tablets, was shared throughout the ancient Near East, whose languages it could notate as syllables, forming a literary tradition flowing in all directions as the medium for the enlightenment and entertainment of those times. Noah's predecessor, Utnapishtim, made an ark out of his straw house when warned that the gods would destroy mankind. This ark was a cube of side length 60 fathoms, a fathom being 5 feet. Its volume was therefore 216,000 cubic fathoms, or 27,000,000 cubic feet, which can then be seen as 6^6 times 10^6 (60^6) inches.

Noah's Ark was differently specified as 300 cubits long, 50 wide, and 30 high, reflecting perhaps the assimilation by the Jews of Egyptian influences, which began with Moses being placed in a basket* by his mother and floated on the Nile to avoid a cull of young Israelites. Moses was then found by an Egyptian princess who adopts him and he then trains as a royal priest. Escaping with his people, he was also

---

*The Hebrew word for "basket" is the same as the word for "ark," *teba.* It only occurs twice in the Bible, for Noah's Ark and for Moses' basket, in both cases representing salvation from waters.

Figure 7.11. A window featuring the Hebrew tetragrammaton יְהוָה in St. Charles's Church, Vienna, Austria. Note the equilateral triangle indicative of the three dimensions of a cube.

escaping with the knowledge of a high Egyptian priest and therefore of metrology, which then shows up when the Bible meticulously defines the dimensions of the Ark* of the Covenant, Solomon's Temple, and other sacred constructions in cubits.

The God with whom the Jews had made their covenant was YHWH, whose Canaanite letter-numbers are 6.5.10.5. Ernest McClain sees a code here for six to the power of five times ten to the power five and thus 60^5 represents YHWH.[7] We have seen above that the cubit used at Jerusalem was probably the same unit used within Noah's Ark, a cubit that is one-thousandth of 12^3 (twelve cubed over ten cubed) or 1.728 feet. When applied to Noah's Ark of 300 by 50 by 30 cubits, this gives 777,600 cubic feet, exactly one thousandth of 60^5, a power

---

*The "ark" in "Ark of the Covenant" is represented by a different Hebrew word from that for "basket" or the "ark" of Noah.

and symbolism not present within the 60^6 cubic geometry in Utnapish-tim's Ark. We can also note that Noah's Ark, described in the first millennium BCE, is using the foot-based metrology of the ancient world, while the Sumerian Ark used cubic inches, the measure used for day-inch counting since Carnac in the fifth millennium.

If one turns back to the Earth and its dimensions as a primary source for ark building everywhere, we see that the number of inches in the meridian is 1.01376 times 60^5 so that the geographical inch of 1.01376 inches reveals the meridian as being a geographical YHWH of 60^5 or 777,600,000. This ties in perfectly with Jerusalem's Temple Mount walls revealing the length of the mean Earth circumference as 25,920 Greek miles, the number of years in the precessional cycle.

The last book of the Christian Bible's New Testament, *The Revelations of St. John of Patmos*, describes a cubic New Jerusalem with

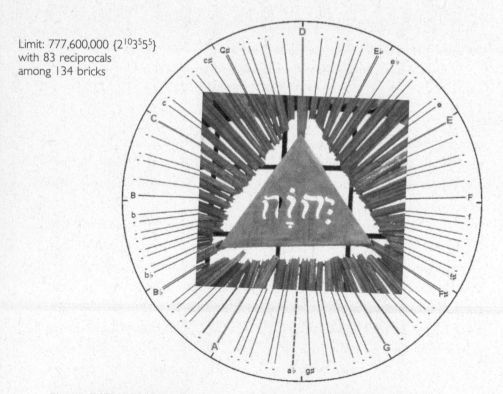

Limit: 777,600,000 $\{2^{10}3^55^5\}$
with 83 reciprocals
among 134 bricks

Figure 7.12. YHWH represented as the harmonic potential of 777,600,000, emanating from the triangle where each side equals the cube root of two.

side length 12,000 fathoms (60,000 feet), which cubed gives a volume of 1.728 times 10^{12} or 10 million million Jerusalem cubits. If, as John Michell suggests, this side length is a macrocosm of the idealized mean Earth diameter (held within a cube), then 7,920 miles divided by 60,000 (feet) gives a furlong of 660 feet of 1.056 feet (696.96 English feet, because 5,000 such feet equal the English mile of 5,280 feet). This New Jerusalem, the mean Earth, was a cube constructed to contain the spherical mean Earth, then connecting to the ancient literary tradition of ark building and its metrological context of cubes related to the mean Earth.

This reveals the nature of such flood events as taking place in the cultural mind of human beings. Indeed, this flooding phenomenon is referenced within myth as the history-making environment of the planet. It acts upon the cultural mind to generate epochs of human development punctuated by a tyranny that threatens the traditional notions of there being any spiritual purpose to human life. The mean Earth, then, represented a world of spiritual values intervening within the human world to bring about creative changes.

The significance of Jerusalem and of the meridian to a covenant with YHWH (as 777,600,000 geographical inches) would point to a harmonic god of latitude (see figure 7.12), quite different from the equator, which is simply circular and ruled by the sun that rises and sets every day. Destiny was seen set at right angles to Fate, on the celestial sphere.

Destiny is not an everyday affair and, looking at the divine numbers of the Greek god Apollo, whose oracle was located at Delphi, we find Apollo associated with 60,* 6^5 (7,776),† and 60^5 (777,600,000). Thus we might assume that Apollo, known as a god of the Earth and

---

*Said to be "In some ways like its father (three) and in some ways like its mother (two)" because the product of 3, 4, and 5 is 60 or 6 times 10 where 3 times 2 equals 6 and 2 times 5 equals 10. However, reading McClain under "Plutarch": "Plutarch (c. 50–120 CE) has handed us a key [in *Isis and Osiris*, 373–374] not only to Plato's mathematical allegories but probably also to the numerology in the mythologies of several older civilizations."[8]

†Plato's essay *Ion* succeeds in answering the riddle "But where is Apollo in *Ion*?" by giving the essay 7,776 syllables.[9]

of riddles about the future, held a similar meaning as YHWH for the Greeks visiting that most special oracular center Delphi, whose latitude was 3/7 of the northern hemisphere and therefore between Karnac at 2/7 north and Avebury at 4/7 north.

Jerusalem provided processional continuity for the figure of Jesus and this led to Jerusalem being seen as the center of the world for Christianity and Islam, just as Delphi had been for the Greek world. While a simple expression of local space sees the horizon and cardinal directions as structuring the world for the potentially sacred establishment of a center, the precessional structuring of Jerusalem by YHWH and of Delphi by Apollo indicates a much more sophisticated sense of the sacred at work and of a continuing sacred history. It is possible that the sense of a precessional god can guide us to understanding what has been happening to humanity since the Stone Age.

8

DESIGNER PLANET

The universe begins to look
more like a great thought
than a great machine.
                    SIR JAMES JEANS, 1930

THE MONUMENTS OF the megalithic era have so far been interpreted
as necessary technical designs, suitable for understanding astronomical time periods and for establishing the dimensions of the Earth and
distances upon the Earth according to relative longitude or latitudinal degrees. The last chapter ended with developments emerging from
the geodetic phase, such as Jerusalem's Temple Mount perimeter walls
symbolizing the Precession of the Equinoxes, when Jerusalem became
involved in a drama related to the changes between two world ages. We
also saw how Sumerian, Jewish, and Christian texts detail metrological
designs for arks related to cubes and for buildings or shapes whose areas
and volumes generate large numbers related to planetary dimensions
and astronomical time periods, thus providing a link between heaven,
Earth, and humanity's sacred spaces. Such sacred structures remained
significant to those initiated in megalithic metrology, but when presented as religious symbols, they were increasingly detached from any
factual relevance.

Monotheism cut right through all necessary supports for what was

Figure 8.1. Giza Necropolis (above) and Stonehenge (below), manifestations of the same geodetic activity that revealed the dimensions of the Earth in miniature. Built within one century of each other, they use the same metrological system to demonstrate different aspects of the Earth, using the simplest possible numbers and a pattern later used for the foundations of sacred buildings.

sacred by proposing that the one god is the only sacred ruler and that our faith in that god should be the focus of religion. This has been handled in quite different ways: protestants denuding their symbolic landscape with white walls, Muslims spurning representations and preferring patterned decorations, and Catholics populating their churches with religious pictures. But each group, to a lesser or greater extent, borrowed from the original canon for the architecture of religious spaces, in which metrology provides a sacred link to the Earth, if only through the fact that our measures were once used to measure the Earth.

Humanity's cultural development was therefore established upon a foundation provided by the astronomers and geometers of the megalithic, but when built upon, this megalithic foundation was soon forgotten and lost sight of. The combination of number and symbol within the prehistoric period was projected into historical cultural settings as a canon of sacred geometrical art. This symbolic work turned megalithic understandings into simple and powerful patterns, such as the ark, which the ancient world reset into traditional information (texts and stories) for transmission across unknown future times and circumstances or within temple-building practices. These stories were largely formulated within

special scribal centers, versed so to speak in the evolving number sciences of the East, incorporating all that was known about geometry, art, temple-building, and compositional and harmonic codification.

These new number scientists of the ancient Near East inherited the proto-religious idea that the world was wholly founded on numerical relationships. Carnac's astronomical relationships involving multiple squares indicated how numbers might participate in cosmic structures and rhythms. The shape of the Earth and the role of the moon and planets had then become known through the use of ideal units of length, within a pattern surely not dreamed up by human agency. This pattern of the circular mean Earth was resolved into a specific geometric template, inscribing the Earth within a design and revealing the Earth's unique relationship to the number field. It was this view that rightly turned megalithic science into the ideals we find embodied in sacred religious structures, these being perpetuated in the design of temples throughout the historical period. Temples could be centers of the world within the landscape or enclosed structures built of stone, each numerically coded using a metrological system to establish numeric connections between the Earth and the heavens.

THE GEOMETER'S GIFT

At Carnac the first four numbers were seen to generate multiple-squares and in modern math we sometimes multiply numbers together so as to form factorial numbers such as factorial four, shown as 4!, which is $1 \times 2 \times 3 \times 4 = 24$ in total, as a product. The factorial numbers are easy to play with and they can be arranged in a series as

1!	2!	3!	4!	5!	6!	7!
1	2	6	24	120	**720**	**5040**

By now we know that seven is useful within pi as 22/7, noting then that 22 is two times eleven. If 5,040 were declared as the length of a radius, then it would have a circumference equal to 6! times 44, or 720 times 4 times 11, which equals 31,680. One quarter of this would be the length of a single quadrant equaling 7,920, the idealized diameter of the

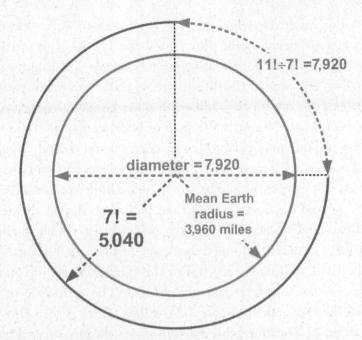

Figure 8.2. The relationship of the mean Earth diameter to the earliest numbers. Shown as 11! divided by 7!, this relationship defines two circles in which the diameter of the inner circle is the same as the quadrant length of the outer circle, which has a perimeter of length 31,680 units.

mean earth in English miles. This length happens to be 11! divided by 7! since 8 × 9 × 10 happens to equal 720, which is 6!. This can be presented as two circles with a common center as in figure 8.2.*

Perhaps the proximity of the inner circle's radius of 3,960 to the actual mean Earth radius of 3,958.69 miles was seen as the prototype of this relationship between the factorials 11! and 7!. These two lengths for the mean Earth are systematically related as 3,025/3,024, a fact that will come up again in chapter 9.

The presence of factorial numbers in late Stone Age thinking should not come as a shock since, throughout the megalithic period, the art of

---

*The geometry involved was intuited by John Michell through his lifetime of studying Stonehenge and the Great Pyramid. What follows in this section is my chosen path through his analysis of it. Please consult his latest *Dimensions of Paradise* (2008) for his treatment.

factorization resulted in whole numbers being seen as the product of their smaller factors, including the prime numbers. The lesser factorials, such as 4! = 24, would be present as aggregate metrological lengths, so that 6! equaling 720, seen in the dimensions of this mean Earth pattern (as common factors within the radius and outer circumference), would be 30 times 24, where 30 equals 5 times 6 or 6! divided by 4!. The resulting geometry is quite unique. No other geometry could embody a quadrant to equal a diameter, engage in such fundamental play of factorials, as well as numerically express the Earth's size in miles. Built in the same century, Stonehenge and the Great Pyramid can be seen to display this relationship of two circles, albeit in completely different ways. This implies that this factorial geometry informed the design of these monuments.

The Earth and moon, so vital to the conditions for life on Earth, are presented as an ideal geometrical form: The difference in radius between the two circles in figure 8.2 is 1,080, which is the radius of the moon in miles. Therefore, this diagram shows the lunar radius between the outer circle and inner circle, and the outer circle therefore indicates the center of the moon (see figure 8.3). John Michell suggests this outer circle probably corresponds to what came to be called in ancient times "the sublunary world," the region of the cosmos from the Earth to the moon, the world below the moon's geocentric "orbit" and all within the sphere of lunar influences.

The addition of the moon to this diagram (shown in figure 8.3) reveals a designed aspect of the Great Pyramid, whose height is seven relative to its base (south side) of eleven. The geometry generates a surprisingly good approximation to the moon's diameter, relative to that of the mean Earth, as being in the ratio of three to eleven. The full height of the Pyramid is 481.09 feet (441 Sumerian feet, each of 12/11 feet) and its southern base length is 756 feet (441 royal feet of 12/7), which multiplied by a height-to-base ratio of 7/11 equals 481.09 feet. The 441 Sumerian foot height could then have one Sumerian foot (1.09 feet) removed to achieve the 480 foot height found in the Pyramid, which then, in parallel, makes the mean Earth radius of 441 units relate to the polar radius length of 440 units, each unit being about nine miles long within the actual Earth. One should contemplate how such tricks were conceived by the megalithic architects of Egypt since it reveals a

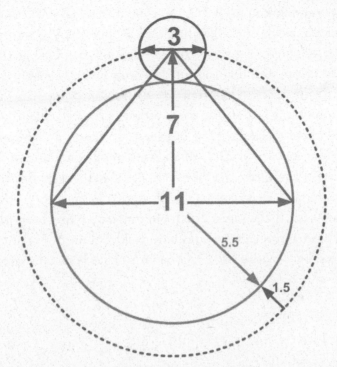

Figure 8.3. The moon (small solid circle at top) is related to the mean Earth (larger solid circle with radius 5.5) as shown within the height to width ratio of the Great Pyramid, which manifests the pi/2 = 11/7 of the mean Earth diameter to the 720 times 7 radius of the "sublunary world" (dotted circle).

watershed between metrology and the later functional mathematics of the Fertile Crescent.

Another remarkable property of these two circles is that the square within which the mean Earth sits has the same perimeter as the outer circle of the sublunary world, thus squaring the circle in terms of its perimeter (if pi = 22/7). The reason is childishly simple: The perimeter of the square is four times eleven or 44, while the larger circle is radius seven times two pi, which at 22/7 also gives a perimeter of 44.

Also of interest is the fact that with a pi of 22/7 in this model, the difference between this square's perimeter (44) and the mean Earth circumference (11 times 22/7) is the circumference of the moon, radius 1,080 miles: thus the moon's circumference is the difference between 4 and pi, when equal to 22/7 (3.<u>142857</u>), multiplied by 11. Again, there is a simple reason: Since pi is 3 1/7, then the difference is 6/7 and multiplied by eleven yields 66/7. Dividing then by pi of 22/7 gives the moon a diameter of three relative to the eleven unit diameter of the mean Earth. The eleven unit diameter of the mean Earth is enabling a rational three for the moon's diameter.

The important fact is that these properties were achieved through the actual three to eleven ratio between the moon and mean Earth, assuming a rational pi of 22/7, to an error of one part in 2,500, or 1.6 miles. This is the difference between the actual mean Earth radius, say 3,958.69, and that assumed within this geometry, 3,960 miles. This means that if the mean Earth circumference had been measured correctly, then a pi of 22/7 would have yielded a mean Earth radius of 3,960 miles, but a pi of 864/275, known to have been employed in the

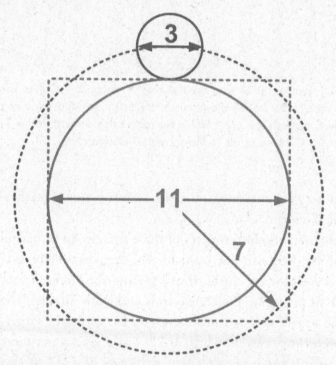

Figure 8.4. The perimeter of the square that encloses the mean Earth diameter of 11 units has a perimeter equal to the sublunary circle of diameter 14 units, leaving the diameter of the moon, 3 units, as the difference between them. Also of note is that the moon of three units has a perimeter equal to the difference between the square perimeter of 44 and the mean Earth circumference. These properties all flow from employing a pi of 22/7.

megalithic, would have yielded 3,958.69. In terms of achieving whole number resolution of a complex geodetic relationship, this geometry gives an excellent summary of the relationship between the Earth and the moon. The earliest known application of it appears to have been in the center of Stonehenge.

John Michell offers advice on how to view this geometrical pattern:

It would be easy to conclude that [this diagram] was actually modelled on the figure of the earth and moon in some remote age when the dimensions of the solar system were reckoned more accurately than was possible in early historical times. Yet the traditional code of number set out in the New Jerusalem* expresses the organization of . . . the inherent framework of number itself, [and] one must be wary in ascribing origins. It is a reality formed by the interplay of creative forces that can most adequately be likened to a pattern of numbers. The squared circle of earth and moon is therefore best described, not as the primary model for the . . . diagram, but as its astronomical expression.[1]

THE PATTERN PLACED IN STONEHENGE

Britain's most unique manifestation of megalithic science was Stonehenge, which may have used the squared circle design (see figure 8.5) and hence itself relate to the mean Earth diameter and also to the fundamental properties of the early numbers up to eleven and a pi equal to 22/7. However, this monument was packed full of additional information, forming a primary reference for the model of the Earth through its pole to mean radii and for the metrological system based upon the Earth. John Michell was the prime exponent of this view.

The temple within Stonehenge (see figure 8.5) consists of two concentric stone circles that enclose two U-shaped stone arrangements.

---

*John Michell anachronistically called the pattern that of "New Jerusalem" because it appears described in the *Revelation of St. John*. This name has been avoided here to prevent confusion. If Stonehenge was an early example of the pattern, it must have been innovated by megalithic scientists but later referred to by St. John of Patmos.

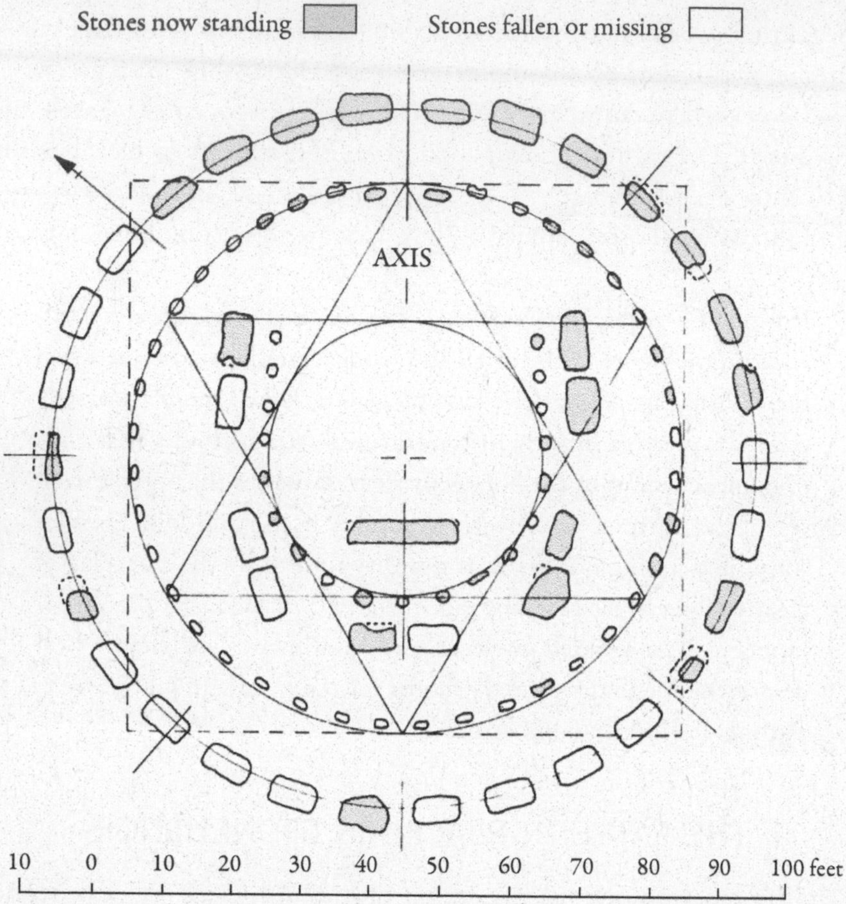
Stones now standing
Stones fallen or missing
AXIS
10 0 10 20 30 40 50 60 70 80 90 100 feet

Figure 8.5. The ground plan of Stonehenge when overlaid by the geometry, which may well have determined the form of the temple. The mean circumference of the Sarsen Circle of lintels is 316.8 feet, and a square with perimeter of 316.8 feet would contain a circle enclosing the bluestone ring, diameter 79.2 feet. The 19 stones of the inner U-shaped structure enclose a circle of diameter 39.6 feet, within a hexad. Illustration by John Michell, from *Dimensions of Paradise*, p. 32, fig. 6.

The outer Stonehenge circle, the Sarsen Circle, consisted of 30 pillars of Wiltshire sarsen stone of which only 17 remain, these supporting a continuous ring of 30 curved lintel stones, now reduced to 6. The lintels were fitted together with tongue-and-groove joints and were held in place by stone tenons protruding from the tops of the pillars to fill holes in the lower surfaces of the lintels (see figure 8.6).[2]

The mean circumference of the lintel ring is 100.8 × 22/7 = 316.8

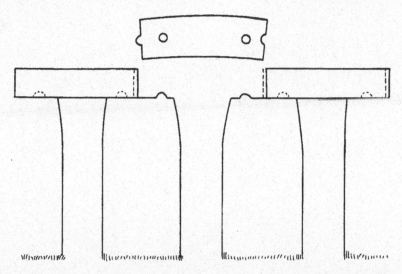

Figure 8.6. The unique Sarsen Circle of Stonehenge in which sculpted stones raised jointed lintels to form a ring. Illustration by John Michell, from *Dimensions of Paradise*, p. 30, fig. 5.

English feet, which directly relates the lintel ring to the foundation pattern detailed above, using dimensions scaled down by one-hundredth of the full pattern, in whole numbers of feet. Such directness of application seems to confirm the existence of this pattern in the minds of the designers of Stonehenge. By using the mean diameter of the ring of sarsens, the ring's outer and inner diameter could be used to carry further information about the Earth.

Figure 8.7 shows the simplest way to see the Sarsen Circle, as three diameters: an outer diameter of 30 royal yards, a mean diameter of 29 royal yards, and an inner diameter of 28 royal yards. These dimensions leave the width of the ring as one royal yard. This is the same size as the width of the Aubrey Circle,* suggesting some form of continuity to build a monument in the center based upon the same thickness of ring and using the same unit of measure, the royal yard, which also indicated that this royal yard was *in use* in 3000 BCE by the builders of Stonehenge 1.

How could this royal yard have been available? We know that the

---

*The average diameter measured for the Aubrey "pits" by archaeological work of Stuart Piggott, Richard Atkinson, and Richard Colt Hoare in 1950 was 1.06 meters.[3] This differs from a royal yard of 3.475749 feet by one part in 1,800.

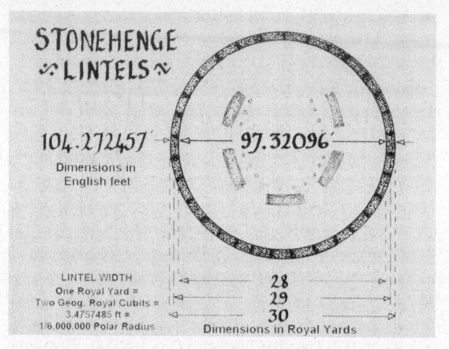

Figure 8.7. The key dimensions of the lintel ring suspended by the Sarsen Circle. Figure by John Michell, from *The Measure of Albion*, fig. 8.1.

root value of the royal yard, 12/7 feet, was marked in Gavrinis around 3500 BCE. A geographical royal yard is two of these, 24/7 feet, and then adjusted by the standard geographical ratio of 1.01376 (3,168/,3125) to become 3.475749 feet long. This geographical ratio makes 360,000 English feet on the equator become the 364,953.6 feet of latitude 51 to 52 degrees, which equals 210,000 geographical royal cubits, and 105,000 royal yards. The geographical ratio (effectively 364,953.6/360,000) appears first identified during Stonehenge 1 as the ratio of 1.056 feet (dividing that latitude as 345,600 such feet) to 25/24 feet (which divides 360,000 to give 345,600 such feet), as shown in chapter 6.

The clear lengths of royal yards in the Sarsen Circle's three diameters hold specific relationships to the mean Earth and polar radius because of the 11/7 ratio found in the geographical royal yard used, namely 3.47575 (24/7 × 3,168/3,125) feet. The builders of the Sarsen Circle already knew that the royal yard divided well into the mean Earth as 189×10^5 and the polar radius as 6×10^5. Their task was to relate the mean Earth to its own radius and the polar radius within a suitable monument to record a key feature of their model of the Earth, the ratio 441/440 (63/20 times 7/22), the product of canceling two different approximations of pi.

The outer lintel diameter of 30 royal yards had the following function: There are 6×10^6 lintel widths (of one royal yard) in the polar radius and 378×10^5 lintel widths in the mean Earth circumference (twice the length of the Earth's meridian).* When these numbers are divided by 30 royal yards, the outer diameter of the lintel ring, the Earth's meridian (half of 378×10^5 royal yards) becomes 63×10^4 units of 30 royal yards and the polar radius becomes 20×10^4 units of 30 royal yards. Therefore, dividing the meridian by the pole gives us pi in the explicit *decimal* form as 63/20 = 3.15.

The inner lintel diameter of 28 royal yards can be explained as relating the Earth's meridian to the mean Earth radius. Remembering that pi equals 22/7 was assumed by the megalithic builders to be close enough to the actual value of pi, the inner diameter of the ring is 4 times 7 royal yards, so that it can stand for the mean Earth radius of 7×12^6 feet long. The 28 royal yards, when divided into seven parts, will each be 4 royal yards long, and each will therefore represent 12^6 English feet in the mean Earth radius.† The "good pi" of 22/7 will then give 22 units of 4 royal yards on the inner circumference of the monument. The inner circumference is therefore the meridian length relative to the mean Earth diameter. In parallel, the scaling factor can be doubled by using the inner radius of the lintel ring, 14 royal yards, so that the inner circumference becomes the mean Earth circumference relative to the mean Earth radius.

In the relatively small Sarsen Circle of raised lintels at Stonehenge, this arrangement encoded the core of the megalithic geodetic model (see figure 5.13 on p. 121). The royal yard was the chosen lintel width because it was and still is a common denominator of the two key dimensions,

---

*The megalithic builders appear to have dealt with powers of ten within such numbers as a single large exponent, as we do today, a natural extension of the need to keep such numbers in manageable forms of two or more factors.

†Four royal yards are twelve royal feet, each then representing 12^5 English feet in the radius.

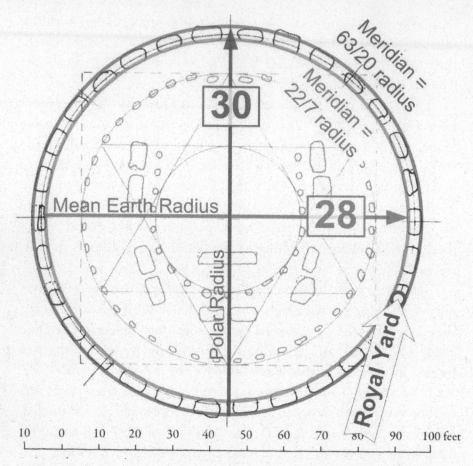

Figure 8.8. The primary design of the Sarsen Circle's ring of elevated lintels, in which the royal yard (the lintel width) enabled the outer and inner diameters to relate the polar and mean Earth radii, respectively, through two approximations of pi, 63/20 for the outer diameter and 22/7 for the inner. Comparing these two circles yields the ratio 441/440, the ratio between mean Earth radius and polar radius, the result of the carefully chosen versions of pi.

the mean and polar radii of the Earth. The polar radius was given the larger diameter *because* it was smaller than the mean Earth radius and hence required the greater pi of 63/20 to express the ratio 441/440 between the two circles.

Through its three circumferences, the lintel ring represented the meridian of the Earth, raised on the 30 sarsen orthostats. We have already discussed how the inner and outer circumferences related to the mean Earth and polar radii. The middle circumference, which runs through each lintel at its half-width point, presents the sublunary world using the 29 full and one half sarsens of the Sarsen Circle to represent the days in a lunar month.

As for its relation to the older part of Stonehenge, Robin Heath has noted the size of the Sarsen Ring as having been set in relation to the Aubrey Circle in the proportion of 19/7, the same ratio that relates the foot to the AMY as 19/7 feet (see figure 8.9).[4]* Since the two rings, Sarsen and Aubrey, have an equal width of 1 royal yard, then 1 foot on the Sarsen Ring's circumference is equal to one AMY on the Aubrey Circle.

LESSONS LEFT IN EGYPT

In the geodetic phase of megalithic activity, the scale of the monuments was large relative to ordinary structures yet was presenting a scaled down model of the whole Earth. This would have been impossible if the Earth had not been found to conform to a number of simplifying numerical rules, such as the variation of pi within the shape of the Earth and the valid approximation to whole number fractions latent within that most ancient artifact in the creation of the Universe, the number field. Whether one takes a view that some demiurge actually visualized such arrangements for the Earth and its moon or simply expects that the number field must have achieved this through resonances within the whole system is in some ways irrelevant. The facts are that the Earth and moon appear to successfully conform to a relatively simple pattern that once noticed by megalithic scientists, enabled manifestation of some of the most ambitious buildings ever built, none more ambitious than the Great Pyramids of Giza.

It seems most likely that the megalithic specialists from northwest Europe chose to integrate into Egypt as a priestly class sometime in

---

*The ring of Aubrey holes has been accurately measured by Thom as being 283.6 feet (+/− two inches), which, divided by the AMY ratio (relating lunar year to solar year over 19 years, using day-inch counting), gives 104.44 AMY. This compares to the outer Sarsen diameter of 104.27 feet with a difference of just 2 inches.

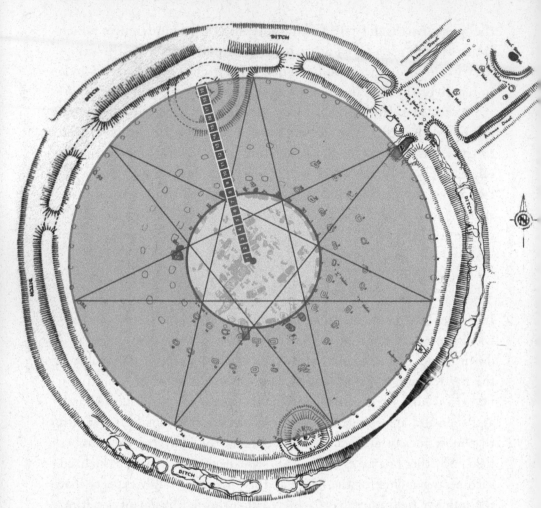

Figure 8.9. The Aubrey Circle has a diameter, relative to the Sarsen, of 19/7, the formula of the AMY in English feet. The Sarsen Circle, 7/19 of the Aubrey, just exceeds the outer crossing points of a sevenfold star. The line of numbered squares shows this relation of seven plus twelve equaling nineteen.

the Archaic period. Perhaps they were associated with the confusion between Horus and Seth briefly giving rise to two types of pharaoh. The first few dynasties must have carried out a geodetic survey of latitudes and came to define Egypt exactly as to its boundaries in latitude and longitude. Then the south would have been surveyed to perfect the simple (but effective) model already obtained between Stonehenge and Avebury, in which the Earth's key radii had been modeled as forming a compound triangle with three lengths of 864, 866, and 867 units between the monuments. At the apex was Stonehenge, which, *after* the work in Egypt, would be enhanced with the Sarsen monuments, the ring of which was made of sculpted stones (uncharacteristic of the British megalithic) and lintels employing tenon and tongue-and-groove masonry similar to the joints often employed in the pyramids of Egypt.

The idea of something large arising metrologically from a small plan can be seen as involving three levels: the properties of the number field, metrological geometrical design, and geodesy. First, the pattern must work in the number field, if metrology is to fashion it, second, into a monument. Third, the monument must then stand for extrapolated measurements of the Earth's dimensions. These three levels can be visualized as forming a triangle, as in figure 8.10.

The relationships found within the Earth, involving pi for example, conform to relationships found to exist in the number field, between specific numbers. Knowing these relationships, as good approximations to the Earth's dimensional variation, enabled metrology to become founded on the number field so that the design of monuments would reflect the Earth, at a specific scaling factor to the actual Earth. Unfortunately, our culture has no theory of numerical conformance, and the megalithic scientists simply accepted what they had found to be the

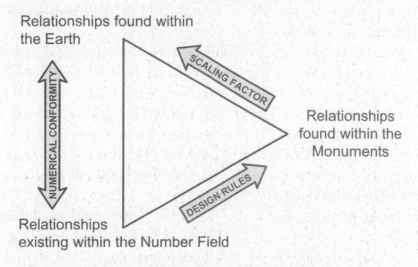

Figure 8.10. The three levels required to understand geodetic monuments.

case. This means that the modern mind cannot believe the proposed structure of ancient metrology and so metrology has been studiously ignored. Any attempts to propose, for example, that the Great Pyramid was a scale model of the earth will only become viable when all three of the above terms are considered properly. And how many specialists today can or will entertain all three to see what the megalithic achieved?

The design rules could emerge from the number field because metrology embodies numeric relationships. We saw in chapter 6 that the three ratios 25/24, 22/21, and 21/20 "gave birth" to the two microvariations, 176/175 and 441/440 (the mean Earth to polar ratio), which have the combined product of 126/125 (1.008) and 3,168/3,125 (1.01376, the *geographical* ratio). These microvariations were then applied to other lengths of feet, which were useful in describing the Earth according to pi, generating whole number perimeters, doubling the volume of a cube, and interrelating different degree lengths. Meanwhile, the English foot was rationally varied to create a range of feet, following the lead of the royal foot of 8/7 feet and whole number variations like 25/24, 22/21, and 21/20. Large ratios such as the cubit of 3/2 were defined as aggregates of a given foot, such as when a royal cubit of 12/7 is 3/2 of a royal foot of 8/7 feet. The full collection of lengths that might be called a foot measure range from 9/10 feet (the Assyrian foot) to 7/6 feet (the Russian foot). Lengths outside this are considered aggregates (like the pygme, pygon, cubit, step, etc.) or subunits (like the digit, inch, knuckle, palm, etc.) of a given foot.

Another fascinating relationship between units of measure and the Earth is found in the relationship between the royal cubit and the cubit one-440th larger. The difference between the two is 12/7 times 1/440, which equals 3/770 (0.0038961) feet, 1/40 of 12/77, the seventh part of the Sumerian foot. When the latter is turned into a geographical value using 3,168/3,125 (1.01375), then the $3,949.\underline{714285} \times 10^{-5}$ foot result is numerically similar to the polar radius of 3,949.$\underline{714285}$ miles, a scaling of 1 foot to 10,000 miles.

In the geodetic period, the megalithic yard evolved rapidly into the new metrological system and came to be based upon the foot of 12/11 feet, a module known to historical metrology as the Sumerian foot. This measure, along with the royal cubit, became crucial in the pyramidion (see figure 8.11) upon which the Great Pyramid was based. The full height of the Great Pyramid (or Pyramid, hereafter) was going to be 441 Sumerian feet of 12/11 feet, for the pyramidion, at 12/11 feet tall, was to express the 441st part, long since lost from the top of the Pyramid.[5] The pyramidion would have been based on the double circles of the foundation pattern (figure 8.3), which could transform the pyramidion's height of one Sumerian foot into a royal cubit through the width of its base, as in figure 8.12 on page 216.

Any rectangle with sides seven and eleven holds this property of transforming Sumerian feet on the shorter side into royal feet on the longer side. This ratio of 11/7 is the quarter section of a circle's perimeter

Figure 8.11. The restored pyramidion belonging to the Red Pyramid of Pharaoh Snoferu, at Dahshur, is now on permanent open air display beside the pyramid it was apparently intended to surmount. Photo from Wikimedia.

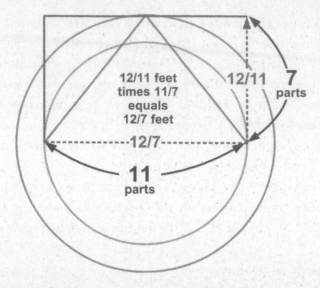

Figure 8.12. Michell's foundation pattern applied to the pyramidion, demonstrating the geometrical form required to transform Sumerian feet of 12/11 feet into royal feet of 12/7 feet as noted by John Neal and the foundational concept behind the Great Pyramid design.

relative to its radius, using pi equal to 22/7, and equates the radius of the mean earth with the area of a quadrant of its northern hemisphere.[6]

This usage of rectangular numbers, resonant of Carnac's multiple squares, becomes clearer in the full Pyramid design that uses a height of 441 Sumerian feet (481.09 feet) to obtain a southern base length of 756 English feet, which in turn equals 441 royal cubits of 12/7 feet. This base length then redefines the type of royal foot employed throughout the Giza complex as being 441/440 of the 12/7 foot royal cubit, of which there are 440 in the southern base length of 756 feet, this then being divisible by the 11 parts in the diameter of the mean Earth (figure 8.12). This organization means that a length of 440 *untransformed* royal yards (of 12/7 feet) will correspond to the truncated height of the Pyramid, which is 440 Sumerian feet. The uncapped Pyramid height expresses the polar radius as being 440 units long relative to the mean Earth radius as 441 units long, while also allowing the southern base to correspond to the mean Earth diameter, the inner circle being the mean Earth. The Pyramid is therefore representing the mean Earth in two contexts: first, through its southern base as the mean Earth diameter and, second, through its height, which touches the outer circle of the sublunary world.

The Earth and the moon were the result of a massive collision of two unequal planets and, while they became a single orbiting system, they were never a single body. Through its dual presentation of the mean Earth, the Pyramid appears to show the moon as having a role in the shaping of the spinning Earth after the collision. This is presented in the re-use of the seven by eleven rectangle to present the degree length north of Giza as a rectangular number, formed by the Pyramid's height and base width: the height of the Pyramid (481.09 feet) times the base (756 feet) generates a product (363,704 feet²) that is within 13 feet of the respected 1907 survey by F. R. Helmert.[7] This measured the degree running south from the northernmost boundary of early dynastic Egypt (the tip of the Nile delta) and was therefore centered upon 31 degrees north. The reducing length of degrees further south then enabled the Pyramid's builders to reduce the base length of the other three sides of the Pyramid to correspond in the same fashion to 30 degrees, 29 degrees, and the septenary degree of Thebes, as was mentioned in chapter 5.*

The problem for the functional mind to grasp is how the Pyramid's geometry could represent, simultaneously, the polar-to-mean manifestation of 440 and 441 (in both its height and width) and the length of a degree of latitude (as a product of height times base width). Such a correspondence implies that the degree lengths within Egypt actually have a relationship to the polar-to-mean radii ratio. This shows that the presence of microvariations within units of measure (including 441/440), the uses of different rational approximations of pi (within the model of the Earth), and measurements of degree lengths all must have been the product of an understanding still blocked today by modern preconceptions about how the Earth was formed through the effectively random condensation of the solar nebula rather than according to the number field.

---

*This was a breakthrough discovery of John Neal since all previous commentators had averaged the base sides thereby eliminating the important differences in length as well as the important information they carry about latitudes.

LOOKING BACK TO EUROPE

In 1992 Robin Cook published an interesting overall plan for the Giza Complex as a four-by-four square with dimensions in royal cubits modified by 441/440, the key ratio built into the Great Pyramid and used throughout Giza.[8] In figure 8.13 the foundation pattern has been superimposed as if it were inscribed within the square of the Giza Complex. If the circle inscribed within the square is Michell's sublunary world, then Khufu's pyramid very nearly marks one corner of a square equal in circumference to the mean Earth circle. Khafre's pyramid is the center of the mean Earth and Menkaure's is centered on the mean Earth circle

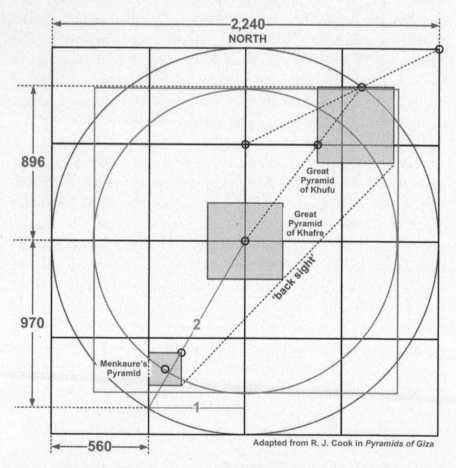

Figure 8.13. The master plan of the Giza complex. Units are in 441/440 x 12/7 royal cubits. Adapted from Robin Cook's diagram in *The Pyramids of Giza*.

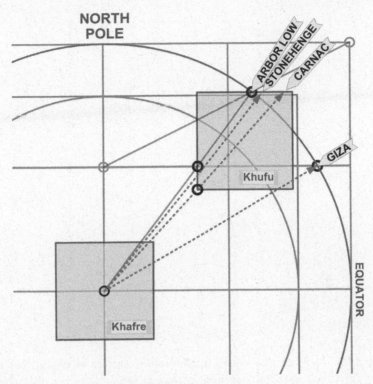

Figure 8.14. The view to the northeast from Khafre's pyramid, as latitudes on the inscribed circle. Referenced by triple crossings are Giza and Arbor Low in England, while our story from Carnac to Britain is shown on the northern face of Khufu's pyramid.

and may indicate the declination of Canopus, the Southern Star, which in 4600 BCE was at 60 degrees south on the celestial Earth, a date corresponding to the construction of Locmariaquer. In any case, the multiple square grid reminds us both of Carnac and of Canevas, as per chapter 2.

If Menkaure's pyramid gives a date through a declination angle for Canopus, then it would seem that the inscribed circle could represent astronomical latitudes upon the celestial meridian, with the Pyramid of Khafre at the center. In figure 8.14 I've illustrated some of these latitudes. The line at 30 degrees north of east represents the latitude of Giza itself, and this crosses the outer circle exactly as the circle crosses the edge of a grid square.* The northeastern corner of Khufu is at 46.2 degrees

*This is a well-known geometrical technique for creating a twelvefold division of an inscribed circle within a square by forming a four-by-four grid within the square.

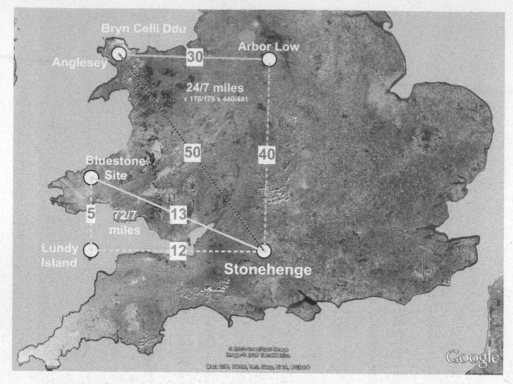

Figure 8.15. The Arbor Low latitude is connected to that of Stonehenge, two degrees further north, through a 3-4-5 geodetic triangle in related units to a 5-12-13 triangle linking Stonehenge to the Welsh source of its bluestones. Base image from Google Earth.

north. The radial line of Carnac, 47.575 degrees, crosses the north base of Khufu's pyramid 38 cubits west of the corner and passes through the southwest corner of the pyramid. The outer circle crosses the north base of the pyramid 186 cubits west of the corner, at an angle to Khafre's pyramid of 53.17 degrees, the latitude of two stone henges called Arbor Low and Bryn Celli Ddu, separated by about 30 units of 24/7 miles.

Robin Heath has pointed to the potential importance of the stone henge called Arbor Low, situated nearly two degrees north of Stonehenge. Its bearing is just over one degree east of north from Stonehenge, at a distance just longer than 40 units of 24/7 miles. A third and important site on the sacred island of Anglesey in Wales forms a geodetic 3-4-5 triangle with Stonehenge and Arbor Low.[9] This complements Robin Heath's discovery of a 5-12-13 triangle connecting Stonehenge with the origin of its bluestones, in the Preseli mountains of Wales, via a right angle at Lundy Island, due west of Stonehenge.[10] The native units of this Bluestone triangle were 72/7 miles, three times that of the Arbor Low triangle's approximately 24/7 mile unit, the same ratio between a foot and a yard. In chapter 6 the Bluestone triangle has a 12 side of 240,000 AMY, showing that the astronomical megalithic yard was part of this unit, since 72/7 miles is 20,000 AMY.

How this unit of 72/7 miles came into existence can be seen by subtracting the polar radius of 3,949.7 miles (3,456 times 8/7 miles) from the idealized radius of the mean Earth, 3,960 miles, the length of the radius in the foundation pattern, to give 72/7 miles as the difference between the radii. Because this unit was employed in the Bluestone triangle, it appears that the foundation pattern was known when the Bluestone triangle was built (around 3100 BCE), using 12 units of 72/7 miles between Lundy Island and Stonehenge. This requires the length of the polar radius to have already been measured before the Stonehenge to Avebury model of the Earth radii was built (chapter 6), so we find, again, that megalithic monuments are often for displaying results (rather than discovering them), making some of what we see today a *retrospective* presentation of previously discovered knowledge.

The geodetic triangle with sides 3-4-5 was more sophisticated in its use of metrology, hence it was constructed around 500 years later, at the same time the Sarsen Ring was being built. Understanding the new units employed gives a missing facet of the system of microvariations.

The surface distance from the center of Stonehenge to the center of Arbor Low is 726,641 feet, but 40 units of 24/7 miles will only reach 724,114 feet. If one divides 726,641 by 40, the result is very close to 24/7 miles times 176/175 times 440/441, which equals 3.44 miles. This unit gives a 40 unit ground distance from Stonehenge to Arbor Low of 726,600 feet, which is effectively exact. The ratio between 24/7 and the unit used equals that between the polar radius of the Earth and the equatorial radius as follows: 3,949.7 miles times 440/441 times 176/175 equals 3,963.3 miles, a very good result. This complements the ratio 441/440, which, on its own, is the ratio of the mean Earth radius to polar radius. The grid of microvariations therefore incorporates both the ratios placed into the model of the Earth in the quarter degree between Stonehenge and Avebury.

The whole triangle was rotated by just over a degree to the east so that Bryn Celli Ddu would land in Anglesey Island. As with the 40

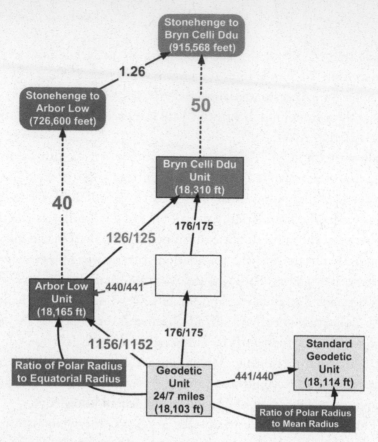

Figure 8.16. The sophisticated use of microvariations in the geodetic 3-4-5 triangle between Arbor Low, Bryn Celli Ddu, and Stonehenge.

unit distance between Stonehenge and Arbor Low, a slightly larger unit was used in laying out the 50 unit distance from Stonehenge to Bryn Celli Ddu. The units used in the 30 unit distance between Arbor Low and Bryn Celli Ddu were increased by the combined microvariation of 126/125 (see figure 8.16).

This microvariation on the 3 side enlarged the 5 side by 1.26, so that Stonehenge to Bryn Celli Ddu, relative to the 4 side, then had the ratio required to double the volume of a cube. We learn from this how the 3-4-5 triangle, considered the fundamental triangle by Pharaonic Egypt, is ideal for achieving a doubling in volume, since the 5 side is already 1.25 the length of the 4 side, requiring only an enlargement, familiar in ancient number science as one part in 125, to double the volume of a cube.

The site of Bryn Celli Ddu was therefore chosen to signify Angle-

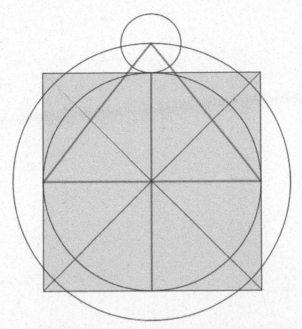

Figure 8.17. The natural integration of the Great Pyramid,
having a four-square base and a relationship between the
mean and polar radius of 385/384 units of 72/7 miles.

sey as a holy island, a tradition that then continued into Druid times
and used the geometric symbolism of a doubling of the cubic altar of
Apollo. This reinforces the identification of Stonehenge as a temple of
Apollo, the geomantic god of Delphi, written about by Diodorus in his
Histories. This also supports our interpretation of Giza, where a range
of latitudes was notated pointing to the locations where the megalithic
work took place, from Carnac to Arbor Low, between Brittany and
southern Britain. The work at Giza could have started in northwest
Britain and returned there, with a fuller metrological system, to found
Druidism in the new center of Anglesey.

Our final motif is to show the foundation pattern as the original reason for pyramid building, with the height of the pyramid expressing a
relationship of the mean Earth to the pole as being 385 units of 72/7
miles to 384 units of 72/7 miles. That is, their ratio is 385/384, which,
times 3, is 1,152 to 1,155, a model for the Earth used in Mexico three
thousand years later.

Figure 9.1. The sacred precinct of the city of Teotihuacan, established 30 miles northeast of Mexico City around 100 BCE.

Figure 9.2. The Aztec calendar stone known as the Stone of the Sun(s) or Five Eras, a large monolithic sculpture excavated in Mexico City in 1790. Approximately 12 feet across, it weighs 24 tons.

9

THE TIME FACTORING OF THE EARTH

ONCE MEGALITHIC KNOWLEDGE had been transmitted to the ancient Near East, it went underground or was absorbed into later cultures. How this can happen is little understood today unless one studies the unexpected role of guilds in the ancient Near East. Social codes, such as that of Babylonian king Hammurapi (1792–1750 BCE), defined all sorts of differences between people and how they were to be organized. Where skilled work was involved, guilds were usually established and these were often multi-national and largely recruited from father to son.[1] It is quite probable that guilds such as architect masons would protect and preserve the original knowledge so that from time to time, an inexplicable upwelling of neo-megalithism would create enigmatic masterpieces, as was the case with the Parthenon built in Classical Athens or Gothic cathedrals built in the Middle Ages. How such artisans "come from nowhere" is the subject of esoteric speculation, but one can recognize within their works a shared corpus of links, to the model of the Earth, sacred canons in art and architecture, and distinctive symbolic ideas and astronomical codes, as if some part of the megalithic was still running underground, like the underground rivers Arcadia and Saraswati in Indo-European myth.

THE START OF A MESOAMERICAN
MEGALITHIC ERA

So it was that beyond the boundaries of the old world culture, in Southeast Asia and Mesoamerica, large stone structures rose again. One of the most extraordinary is Teotihuacan in Mexico, which appeared quite mysteriously around the start of the Current Era, its later Aztec name of Teotihuachan meaning "where man met the gods."

The megalithic technique of multiple squares can be seen at Teotihuacan. Its ceremonial precinct is bounded by a triple-square rectangle that contains the two largest pyramids in the Americas, thought to represent the sun and the moon. The pyramids both lie on the triple-square's diagonal and might well have been associated with the 13-month lunar year, the eclipse year, and the lunar nodal period. The rectangle was aligned to the maximum eastern azimuth of the star Thuban around 170 BCE. This had been the pole star when the Great Pyramid was built at Giza in 2560 BCE.

Nearly three thousand years later, Thuban was "knocked off its perch" as pole star and became one of the bright circumpolar stars. Above the northern horizon, circumpolar stars normally circle an empty north pole, traveling between an extreme eastern and western azimuth with each rotation of our planet, generating northern alignments at sites used for circumpolar astronomy at Carnac in the fifth millennium BCE. It is therefore Teotihuacan's alignment to Thuban, 15.5 degrees east of north, that dates the monument as first conceived in 170 BCE, then adding weight to arguments that the Mexican pyramids were derived in some way from Egyptian pyramid-building using megalithic metrology, multiple squares, astronomical alignments, and, probably, dimensions scaled down from the those of the Earth itself.

The monuments at Teotihuacan anticipated a later Mesoamerican

Figure 9.3. (Opposite) The outline of the sacred precinct of Teotihuacan is a triple-square that relates the thirteen-month lunar year, the solar year, and the eclipse year. Of the three major buildings, two are pyramids on the diagonal of the triple-square. The courts, road, and plaza terminate in the Pyramid of the Moon, all aligned (with the triple-square) to the circumpolar star Thuban, which was the pole star when the Great Pyramid of Giza was built.

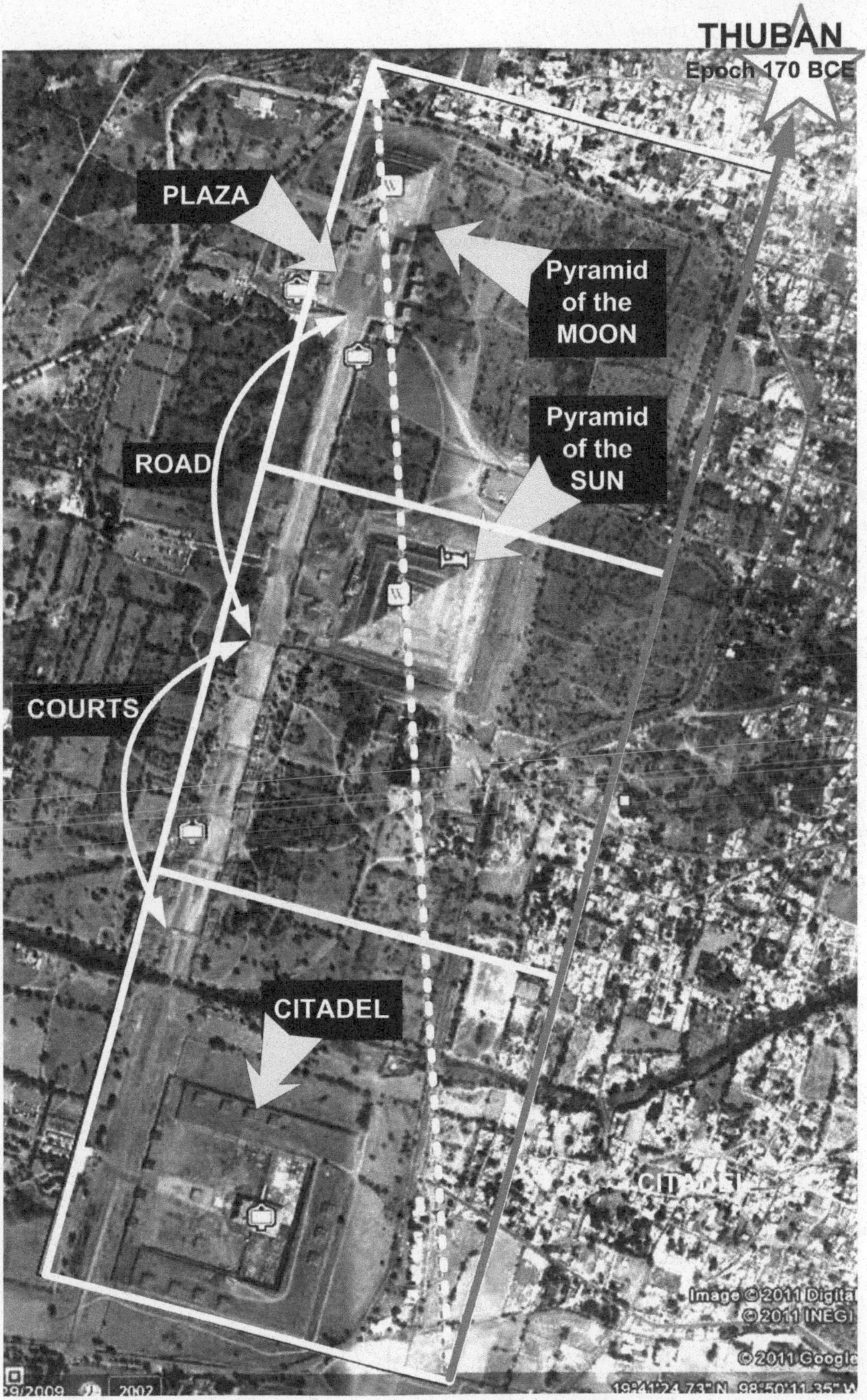

THUBAN
Epoch 170 BCE

PLAZA

Pyramid of the MOON

Pyramid of the SUN

ROAD

COURTS

CITADEL

CITADEL

Image ©2011 Digital
©2011 INEGI
©2011 Google

29/2009   2002

19°41'24.73" N 98°50'11.35" W

and South American pyramid-building period that would use the same sort of astronomical and monumental "technologies." Their use of old world metrology makes it implausible that their construction was wholly independent of old world megalithism, since an original metrology would not have achieved the sophisticated relationship to the size of the Earth found at Teotihuacan. When Teotihuacan and other new world monuments are surveyed, the only metrology found is that familiar at Giza and at Stonehenge, the origins of which are clearly the megalithic people of northwestern Europe, dating back to 5000 BCE.

A ROAD WITH ONE MEASURE AND TWO MEANINGS

Hugh Harleston, Jr., researched Teotihuacan from the 1970s to the 1990s and identified the common unit of measure employed in its design, a unit of about 1.05945 meters long, which he called the hunab. Correspondence with John Neal resulted in the hunab's identification as an old world unit 1.05941 meters long, the royal double cubit, i.e., a royal yard (of 24/7 feet) in its standard geographical microvariation of 24/7 times 1.01376 feet (3.47575 feet), the unit also to be found within Stonehenge and in the Royal module of Egyptian measure used in the Pyramid Age.

Due to my work at Carnac, I soon identified the triple-square as a primary design construct at Teotihuacan, implicit in Harleston's interpretation (see figure 9.4), as having a unit side length of 756 hunabs (a length also found elsewhere at the site). There was a road-like structure running parallel with the sides of the triple-square, this leading directly to the Pyramid of the Moon. Like the triple-square the road would have been aligned to the star Thuban.

The combined courts and road have a length of 1,152 hunabs when measured between their distinct end points. This number is connected to the old world models of the Earth, being the numerical value for the size of the Earth's polar radius in units of 24/7 miles, related to the yard of the Egyptian royal foot, 24/7 feet. That is, there are 1,152 times 24/7 miles in the polar radius. If the hunab were a royal yard of 24/7 feet times the standard geographic ratio of 1.01376, then the road would have been a scale model of the polar radius, where one hunab represented

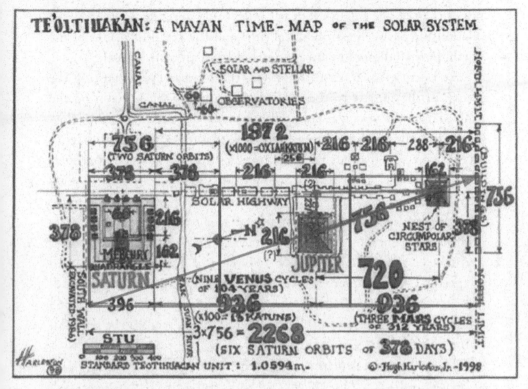

Figure 9.4. Hugh Harleston Jr.'s chart of Teotihuacan as a Time Map, with triple-square added and diagonal that links the pyramids of the Sun and Moon. The courts and road that many call the Path of the Dead he calls a Solar Highway. Harleston's "nest of circumpolar stars" intuits the major alignment of the triple-square and road to circumpolar star Thuban around 170 BCE. Used with permission.

a geodetic unit. However, why was the root royal yard of 24/7 feet not used? In that case each foot on the road would have equaled exactly one mile of the polar radius. Instead each geographical foot (1.01376 feet) in the hunab corresponded to an English mile in the polar radius.

The length of 1,152 suggested a monument where the polar radius was shown as a scaled down length, while the triple-square surrounding it was a megalithic expression of the time relationship between the eclipse year (the triple base length) and the solar year (the diagonal length). This suggested that the lengths of these and other features, like the road, might correspond with astronomical time as well as the dimensions of the Earth. The use of a scaling of 1.01376 feet per mile, rather than of one English foot per mile, came to reveal an unexpected relationship between astronomical time and the key dimensions of the Earth, never indicated in any monument from an earlier epoch.

At Locmariaquer, east of Carnac, the 33-year cycle marking the return of the sunrise to the exact same point on the horizon on the same day of the year was counted out using day-inch counting, traveling directly north to Mane Lud (see chapter 3). This recurrence has become synonymous with the mythic lifetime of archetypal solar heroes, who die, according to many traditional sources, at 33 years of age. The most well-known examples of these solar heros are Jesus, Mithras, and Krishna. The length of the road at Teotihuacan is around four times longer than Locmariaquer's 33-year day-inch count and hence could contain around 132 years of day-inch counting. In chapter 3 Locmariaquer was seen as demonstrating a 33-year periodicity caused by the fractional excess of the solar year length over 365 days, 32/132 (0.<u>24</u>). The prime number eleven is seen in the denominator, times twelve.

The following calculation, in day-inches, can then result:

$$\frac{365.2422 \times 33 \times 4}{41.7} = 1155.913 \; hunabs \; of \; 41.7 \; inches$$

But 1,155.913 hunabs is 3,963.13151 feet, which (in a scaling of one foot to one mile) is the equatorial radius of the Earth in its modern estimate, more accurate than the 3,963.4-mile length of the equatorial radius according to the megalithic model of the Earth.

This calculation can be reversed to find the period that might then be behind the 1,152-hunab length. In other words, if 1,155.913 equates to the solar year, what time period might 1,152 hunabs refer to? When 1,152 hunabs (48,048.75 inches) is divided by 132 (both 33 and 4), an unfamiliar year of 364 day-inches results. This 364-day year had 13 "months" of 28 days (of 4 7-day weeks), which can be reversed as 28 weeks of 13 days. This is the key to this length, for 13 days is the length of the week adopted by the Maya, of which 20 weeks make up the other factor in their Tzolkin of 260 days. The metrology of foot-based measures, used in Britain and Egypt, had moved to Mexico where it reveals a formerly unknown relationship to the Earth, encoded in day-inches.

The 12 inches in a foot enable $4 \times 33 \times 365.2422$ inches to equal 11×365.242 *feet*. That is, when this length was measured in feet, it was 11 times 365.2422 English feet long. When divided by the hunab's standard geographical ratio of 1.01376 (3,168/3,125), the 11 becomes 10.85069 (3,125/288). Teotihuacan's use of the hunab, which was a geographical royal yard rather than a root Egyptian royal yard of 24/7 English feet long, reveals a time-factoring relationship between the radial dimensions of the Earth and different year lengths. The hunab's provision of the required geographical ratio in the road's length indicated a parallel relationship to what is called the Saturnian year of 364 day-inches.

THE GEODETIC LINKS TO DAY-INCH COUNTING

The hunab's relationship to geodesy becomes clearer when we look at the third radius, the mean Earth radius. Repeating the above process *does not* then produce a significant time period. Instead, the idealized length of 3,960 miles for the mean Earth can be divided by 11 to give 360 days. Such a difference might be expected since the mean Earth is a *virtual* dimension, which could only exist if the Earth stopped spinning, while the geographical ratio has only come into being through the rotation of the Earth. Three hundred sixty days is a year familiar to many past cultures, including the Egyptians. In Egypt in the custom of having a year with 5 days left over, those 5 leftover days were called *Neters* in Pharaonic Egypt. Some recent authors have interpreted mythic reference to 360 days as a past year length, prior to some natural disaster that (again) increased the planet's rate of spin. It seems likely that an idealized 360-day year was the time equivalent of an idealized Earth radius of 3,960 miles, as per the foundation pattern. Indeed, an 1,152-hunab model of the polar radius leads to an 1,155-hunab length for the mean radius, which gives the mean radius as 3,960 miles (the common unit of 24/7 miles times 1,155 equaling 3,960).

The mean Earth radius (1,155 hunabs) is then exceeded by the equatorial radius (1,156 hunabs), and the polar radius (1,152 hunabs) is exceeded by the mean Earth radius (1,155 hunabs). From the calculations above, the variation (through spin) of the polar and equatorial radii, from the mean radius, has occurred according to some kind of numerical scheme in which the ideal starting point was a mean of 3,960 miles and a solar year of 360 days. The ideal mean radius stands as one-384th part larger than the polar radius, while the equatorial radius stands one-288th part greater than the pole.

So why should time and Earth dimensions come to be metrologically linked through the division of a length (in miles) by 11? I would suggest this has to do with the foundation pattern, where the mean Earth radius is 11 units (i.e., 3,960 miles divided by 11 to give an inner unit of 360, half of 6!). When the polar and equatorial radii are adjusted this must employ 11 divided by 1.01376 (3,168/3,125), giving measurements of 364 and 365.2422. Indeed, designs called pecked crosses, very similar to the foundation pattern of two circles, one smaller one set within the other, can be found at Teotihuacan and around Mesoamerica.

These pecked crosses were studied in 1978 by Anthony Aveni, drawn by H. Hartung, and the research about them later condensed as an appendix of Aveni's *Skywatchers of Ancient Mexico*.[2] There is variation in the exact number of marks within each part of these designs, but the design and its proportions are generally coherent considering they were laid out to fill each part of the pattern with a given number of dots (see figure 9.5). There appears to have been a counting function for days as marks since these pecked crosses often contain 260 peck marks, the number of days in the 260-day Tzolkin period. This period somewhat displaced the solar year count because, within the tropics, the solar year is less significant (the sun traveling directly overhead twice per year and the seasons less distinguished). The inner circle radius is often eleven marks like the foundation pattern, while the outer circle is separated by a further three to five units, where three would be appropriate to obtain radii of 11 and 14. Such similarities to the 11-14 squared circle design grow stronger when the pattern is superimposed upon the Dendera Zodiac from Ptolemaic Egypt and the Aztec Stone of the Five Ages as in figure 9.6 on page 234.

The pecked pattern looks as though it sought to retain the foundation pattern as a set of lines having a number of dots in lieu of the lengths of lines and sectors of circles. The number of dots appears to have been increased, though, in order to count 260 days, rather like

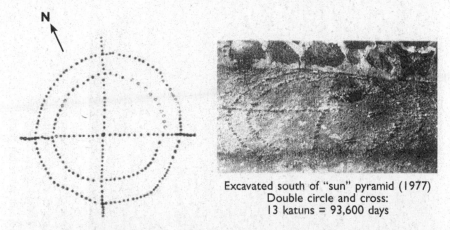

Excavated south of "sun" pyramid (1977)
Double circle and cross:
13 katuns = 93,600 days

Figure 9.5. Two of the pecked crosses found around Teotihuacan and Mesoamerica between 10 degrees north and the Tropic of Cancer. Left: illustration from Anthony Aveni's *Skywatchers of Ancient Mexico*; right: image from Harleston's website.

a day-inch count. The count still preserves the general features of the pattern even though the exact 11 to 14 ratio of the foundation pattern appears lost as an exact geometry.

The Dendera design, called "the only complete map that we have of an ancient sky" by John H. Rogers,[3] has been dated to 50 BCE and is interpreted as being a view of the sky with the celestial north pole in the center. The Aztec design shows Saturnian skull-like divinity at its center, a divinity who controls the five ages. The vertical line in the cross is the polar axis of the Earth and the horizontal is the equator, as in chapter 7's conceptual model of Great Time shown in figure 7.4 (p. 181).

This brings our attention to the unfamiliar time period of 364 days associated with the radius of the north pole. It has been proposed by Robert Graves and others that this Saturnian year was widely applied in matriarchal societies of the past, in which the male king was in some fashion sacrificed after "one year [of 364 days] and a day,"[4] This calendar was probably recognized in megalithic times as it has very good connections to astronomical time, including the synods of Saturn (54 × 7 days) and Jupiter (57 × 7 days), the lunar year (15/16 of Saturn synod), and our present 7-day week. It also restores an important missing link to the megalithic model of the world.

Figure 9.6. Comparison of Dendera Zodiac, now in the Louvre, and the Aztec Stone of the Five Ages, both with the foundation pattern overlaid. On the left one sees punctuation in eight places in both overlay and Dendera border. On the right one sees punctuation in four places, as with the pecked cross designs, and also that the lunar diameter is shown in the pecked design as an extension beyond the outer circle.

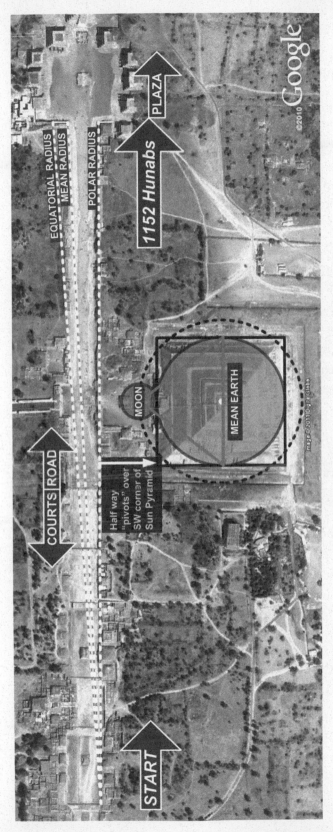

Figure 9.7. The road and courts of Teotihuacan, representing the polar radius, with the other two Earth radii shown as a right triangle. The polar length of 1,152, a harmonic number, has a half-length above the southwest edge of the large Pyramid of the Sun (center). In this model the mean Earth radius is 1,155 hunabs long, which points, through II, to the foundation pattern radius of 3,960 miles. The Pyramid of the Sun is here shown as the squared circle of the mean Earth and the moon shown in the Sun Pyramid's plaza. Base image from Google Earth.

SATURN'S CONNECTION WITH THE POLE

Before time began Hesiod's theogony presents Saturn/Cronos as castrating the sky god Ouranos. Saturn thus separated heaven and Earth by means of his fateful sickle, that is, after he had established the obliquity of the ecliptic, according to *Hamlet's Mill*. The tilt of the Earth's axis created a solar year not made up of a whole number of days having grown shorter by less than one minute over 6,000 years, from 365.2425 to 365.2422 days.[5] As mentioned earlier this length can be fractionally rationalized as 365 plus 32/132 days, involving 11 but not 7 days or 13 days, the week lengths of the old world and new world, respectively. In contrast, the Saturnian year of 364 days could be divided by a whole number of both these types of week.

The reason for titling this *artificial* year length in honor of Saturn, a planet on the ecliptic, while also relating it to the length of the polar axis (as above) is that the synod of Saturn (54 weeks of 7 days) divides evenly by 7 days, resulting in 52 weeks of 7 days.

Saturn is the outermost visible planet and was therefore considered the limit of planetary creation, associated with boundaries and with measurements that cut things up. This started when he cut up Ouranos, the sky god, to create time, and he easily takes the part of Seth, brother of Horus, who cut up his sky-god father Osiris. In Sumerian myth the word *Shamash* conflates sun with Saturn, and this may be due to the twin principles operating between the solar equatorial world of Time and the saturnian polar world of Great Time, or Eternity. According to the myths of many regions and different times, the world was anciently conceived as a great mill, once made by a great Smith but whose operation had been set out of kilter by some ill deed.

It is no idle fancy that the representative of the celestial smith, the King, is himself frequently titled "Smith." Jenghiz Khan had the title "Smith." The Chinese mythical emperors Huang-ti and Yu are such unmistakable smiths and Huang-ti, the Yellow Emperor, is acknowledged to be Saturn. And just as the Persian Shahs held their royal jubilee festival after having reigned thirty years, which is the saturnian [orbital] revolution, so the Egyptian Pharaoh also celebrated his jubilee after thirty years, true to the "inventor" of this festival, Ptah, who is the Egyptian Saturn, and also Deus Faber.[6]

As we "heard" in chapter 5, in the section "Coincidences as Relics of the Creation,"

Kronos-Saturn has been and remains the one who owns the "inch scale," who gives the measures, continuously, because he is "the originator of times," as Macrobius says, although the poor man mistakes him for the sun for this very reason.[7]

The alignment of the Earth's polar axis is directly linked to the planetary world through gravitational forces acting upon the spinning Earth. There is an ever-varying precessional force, which slowly moves the axis of the Earth along a systematic path, then causing the equinoctial points (where ecliptic crosses the celestial equator) to move retrograde. Thus "the long term dynamics of the planetary system is the dynamics of gravitational resonances,"[8] despite the stability found in the lengths of planetary orbital periods. The different synodic periodicities of the planets are generating very low frequency gravitational waves, from the geocentric perspective of each spinning planet. The large angular momentum of the Earth prevents the polar axis from moving its alignment, in the expected fashion, toward another body. Instead, the precessional effect translates forces at right angles to the force applied, so causing an axial drift. The momentary variations in the gravitational forces become aggregated in shifting the polar axis so that the orbital dynamics of the solar system, archaically conceived as "within the orbit of Saturn," are uniquely affecting the alignment of the Earth's pole. While the long-term average of these forces is our familiar Precession of the Equinoxes, this phenomenon is actually made up of short-term vectors that are highly variable, causing a wobbling motion in the polar axis called nutation, due to very low frequency gravitational waves from the planets' motion around the sun. While all of the heavier solar system objects are adding up in a variable way to create a precessional force upon the polar axis of Earth, the sun and moon also cause measurable periodic variations, due to their changing positions. These oscillations, in both precessional speed and axial tilt, are known as the nutation.* Thom shows, in his *Megalithic Lunar Observatories,* that the site called Temple Wood, through backsights to hilltop foresights, enabled prediction of the wobble caused by nutation depending on how high the moon stood on the ridgeline of the hill between 260 days before (the tzolkin) and 173 days (half an eclipse year) before, that is between ½ and ¾ of an eclipse year before, revealing the tzolkin as ¾ of the eclipse year.

This cosmic factor manifests not within the bulge of the tropics, where the Earth is tugged by the sun and moon, but at the pole and it is then under the traditional charter of Saturn, god of time and keeper of limits. This is, I believe, the reason why Saturn appears in the center of the Aztec Lord of the Five Ages and is often represented as having a skull head. He is routinely shown as a skeleton in the old world, holding a scythe, sickle, or shell to cut things into measured parts. Saturn, it seems, is the domain of Time manifested in the tropics as "his" planetary domain and at the pole as planetary gravitational influence over the variation of precessional time.

This allows the planetary astronomy at Carnac to be connected to the action of the planets upon the polar axis. The reason Olmec Teotihuacan might have adopted a 13-day week, as the Maya did thereafter, was probably connected to the eclipse cycle of the sun. Thirteen is a factor with seven in the 364-day year, but it has other potentials seen in a new year length called the Tzolkin, of 20 weeks and therefore 260 days. The eclipse year of 346.62 days is four-thirds the length of the Tzolkin (see figure 9.8) and, during the nodal cycle related to the eclipse year, the moon's forces upon the pole go from a maximum to a minimum and back again. At Carnac this lunar maximum was seen to go, during each orbit, further north and south than the sun did at its own yearly maxima north and south. In this respect the moon, already a great influence on precession, reaches maximum effect at standstill through the invisible agency of the lunar nodes.

We can then note that in myth Saturn swallows (or *eclipses?*) his children, in case they overthrow him, just as he deposed his own father,

---

*The nutation of the axis of the Earth was discovered in 1728 by the British astronomer James Bradley, but it was not explained in detail until 20 years later.

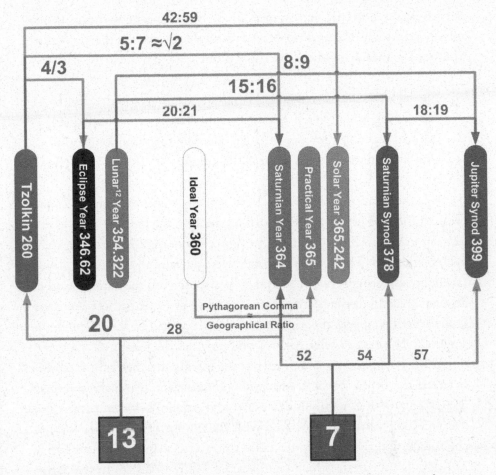

Figure 9.8. Relationships of thirteen- and seven-day weeks to various types of years including the Saturnian year.

the sky. This expresses a fear of change and a willingness to sacrifice even one's own children to avoid it. It could explain the substantial reasons for South American human sacrifice: to appease the forces that brought about the changes they called the Five Ages.

In the Greek version, one son was protected by his mother. This Zeus-Jupiter defeats Cronos-Saturn, liberates his swallowed siblings, and banishes his father's kind, the Titans (that is, the large cosmic beings and the megalithic astronomers who studied them), into various forms of exile where they could no longer influence human affairs with new gods and civil constructs, such as the state and its leadership.

Saturn was reportedly exiled to a small island, perhaps meaning his merely semitone relationship of 16/15 to the moon, or the pole axis, and not the meridian where life is lived out.

> The creator, Auharmazd (Jupiter) produced his creation . . . with the blessing of Unlimited Time (Zurvan akarana) . . . This planet [Saturn] was taken for the one who communicated motion to the Universe and who was, so to speak, its king.[9]

It is the rotation of the Earth that "sets the universe in motion" and creates time, and Jupiter needs Saturn to give him the measure of it, as can be seen in the 18:19 ratio (shown in figure 9.8 on p. 239) between their synodic periods. The relationship between the sun and moon is played out in cycles lasting between 18 and 19 years, between the Saros period of 18 years (plus 10 days) and the Metonic period of 19 years. Between these is the lunar nodal period of 6,800 days (18.618 years). In this the moon is Egyptian Thoth, a god like Saturn who provides the measures.

What does Zeus-Jupiter create but the meridian, which now mediates between the depressed pole and exuberant equator, using the numbers 25,920 and 60^5 (777,600,000, YHWH and Apollo) between Saturn and the sun. Jupiter became associated with the number 12 as the division of the zodiac, and the time Jupiter takes to traverse a single zodiacal sign is 361 days (19 squared). This Jovian year of 361 days relates to the Saturnian year of 364 days in the Gilgamesh ratio of 126/125 or metrological Greek foot of 1.008 feet (to one part in 3,249). This alerts us to a near parallel set of measures within metrology and the time periods of years.

In the region of 360 to 365 days, a numerical difference of one day creates a ratio very like the square root of 176/175 (1.0057), which, as 1.002853, gives a reasonable approximation to 441/440 (1.002272). The difference of three days, between 361 and 364 days, creates a ratio 364/361 (1.0083) very similar to 126/125 (1.008). The difference, overall, between 360 and 365 generates the geographical ratio of 3,168/3,125 (1.01376).*

---

*See the postscript in my book *The Matrix of Creation*, where I discuss these microvariation ratios' role in musical intervals.

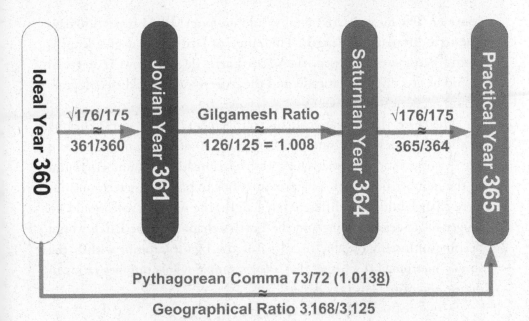

Figure 9.9. The close parallelism between significant time measures acting upon the pole and the microvariations of units of measure in metrology, namely 176/175, 126/125, and 3,168/3,125.

The moon, after the formatory collision with Earth, has been slowing down the rotation of Earth through its gravitational drag upon the planet's water and plastic geological structures. The energy received through this by the moon pushed its orbit further from the Earth, and the moon has now reached a distance of about one-quarter of a million miles. Despite the growing distance between these two bodies, the moon can still, quite magically, when nearest to the Earth just cover the face of the sun during a solar eclipse, making it a total eclipse. Due to the Earth's rotation, the moon's orbital period has reduced from 400 days or more to the present 365¼ days. At the same time, the moon has become resonant with regard to Jupiter and Saturn, their synods now being 9/8 and 16/15 of the lunar year. This situation can only have come about due to the long-term effect of a heavy but distant planet tugging on the moon, when the moon is between the Earth and a gas giant.

The Earth day is very stable but, were it to lengthen, the length of the polar radius would grow and that of the equatorial radius would contract. The mean Earth radius would no longer be 11 times 360 days due to a different day length. The range of latitudinal degree lengths would become compressed, the mean Earth degree (now 51 degrees) would be at a different latitude, and the pattern of all the different years would numerically change.

Therefore, the present geodetic arrangements can be seen to belong to the properties of the numbers between 360 and 365, manifested in the microvariations of metrology and relationships to numbers found in the number field, such as approximations to pi. If these relationships were changed due to a different day length, the number field would no longer relate celestial time and the Earth's shape and the Earth would become unintelligible due to a loss of the facility, found within the simple metrological scheme, that allows simple whole number ratios to define an entire planetary design.

UNDERSTANDING SATURN'S MASTERPIECE

To understand the proposal that numbers, through processes like resonance, impose intelligibility upon the Earth requires us to perform a careful separation of the relationships found in the megalithic model of the Earth into just four types.* These are the relationships between,

1. Numbers
2. Latitudes
3. Years
4. Radii

For the sake of keeping these terms related to each other in our mind, a diagram can be formed in the shape of an interconnected diamond, where one can contemplate how each of these terms has a differ-

---

*Here we borrow from the techniques of Systematics, pioneered by J. G. Bennett in the 1960s and '70s. The four term system is evolved from traditional precedents including Aristotle's four causes. The website systematics.org can provide further information about this technique of holistic multiple term systems including online resources and bibliography.

ent part to play, with respect to the others. The singular key to three of these terms is the spin of the Earth, also called rotation, which cuts up the years into a number of days (as in day-inch counting), lengthens the latitude degree lengths (as one travels north), and increases the difference between the polar and equatorial radii (see figure 9.10).

From the beginning of the megalithic era, numbers were an important influence appearing within time phenomena. Our thesis is that the relationships between numbers came to define the eventual form of the Earth, just as they define astronomical time. The child-like convenience of multiple squares at the latitude of Carnac, a possibly unique center of learning, revealed the extremely organized world of astronomical time and its different years, arranged according to relatively small numbers. The later quest for the size of the Earth started with the equator length, which appears to have been humanly modeled to match year lengths, such that it was 365.242 times 360 times 1,000, containing the ideal

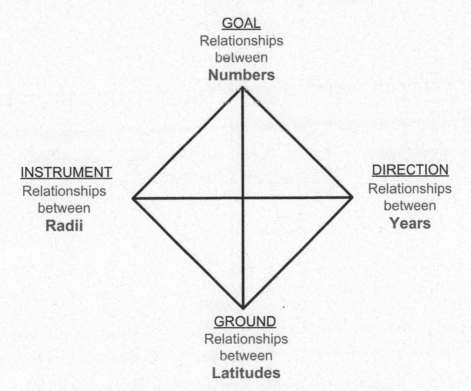

Figure 9.10. The different types of relationship found in the Earth according to its spin.

and actual solar years as numbers and giving a length for the equatorial radius. The search for the mean radius then led to another latitude, that of the mean Earth degree. The cubic relationship of any spinning planet's deformation allowed the polar radius to be deduced, at which point the elliptical figure of the Earth came to be known through numeric ratios between radii.

Following the connections shown in figure 9.11, if one had the "goal" (or if there was a law) that the Earth should conform to number, then the basic "ground" (under our feet) would need to distort with latitude in such a way as to be numerically intelligible. But these two vertical terms, number and latitude, cannot be directly connected; they are simply motivational terms. To have number manifest as latitude, they need to be mediated by operational terms, which require an "instrument" and a "direction," on the horizontal. However, the megalithic scientists found that the principle that links the two was that of

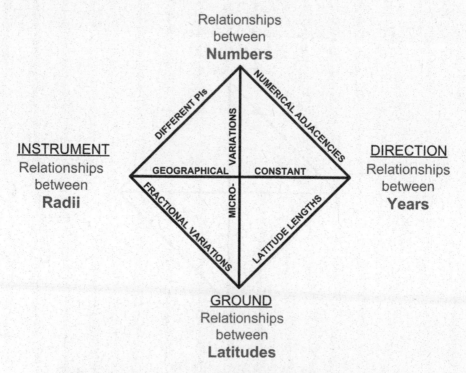

Figure 9.11. The six connections between the four terms enabling contemplation of their relatedness.

the microvariations of the foot. This realizes itself operationally as the geographical constant or ratio of 3,168/3,125 (1.01376), which defines the range of microvariation used within metrology. The "goal" of numbers was highly influential in the use of different approximations of pi, which make up the microvariations through self-cancellation, and in the use of numerical adjacencies involving large numbers such as 176/175 and 441/440 found (approximately) in the relative lengths of year in days (360 to 365) but more exactly in the microvariation of feet that divide into degrees of latitude.

The great "instrument" of metrology is its fractional variation of the foot to obtain compatible divisibility of radii and of latitude. A prime example of this is how the royal foot of 8/7 feet divides into the polar radius so that 12/11 feet can then relate to royal feet in the height to width ratio of the Great Pyramid and the seven to eleven ratio found in the foundation pattern. Such feet, relative to the English foot, are clearly seen in the length of the equator but are less obvious in the mean Earth radius of seven times 12^6 feet, demonstrating that the foot length belongs to the spherical volume of the Earth itself, a fact related to cubit Arks using powers of 12 and 60.

Finally, the "direction" of these arrangements can be found in the connection between the lengths of years and the latitude lengths. By achieving a relationship between the time environment of years and the variation of latitude length on the meridian, the earth appears to carry to fulfillment the vertical, or motivational, pair of terms (number and latitude) through the necessary horizontal, or operational, terms (radii and years).

The microvariations involving different values of pi can now be resolved as a network of relationships constituting a design for the planet "on the theme of pi" as shown in figure 9.13 on p. 247. This could only be seen after the Arbor Low to Stonehenge distance was found to express the ratio of polar to equatorial radius of the Earth and the connection between radii and time, indicated at Teotihuacan.

The system of metrology replaced any need to display models of the Earth with units in fractional proportion, these relating to radii as well as latitudes. These harmonious conditions only came into existence due to the collision that created our moon and the consequent tilt of the

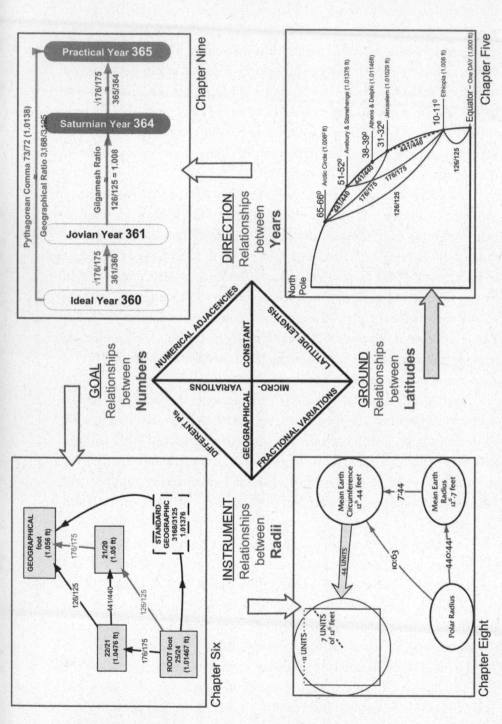

Figure 9.12. Expansion of the tetrad of Earth's form, connecting earlier figures and their connections as harmonized terms.

Earth and its year length in days. In the modern world we may have highly technical ways of computing astronomical and geographical measures, but the ancient scientists of the megalithic discovered the highest achievement in elegant simplicity relating to our world and the heavens above: three radii, two ratios, and one foot, divided in various ways by ratio related to three different pi approximations.

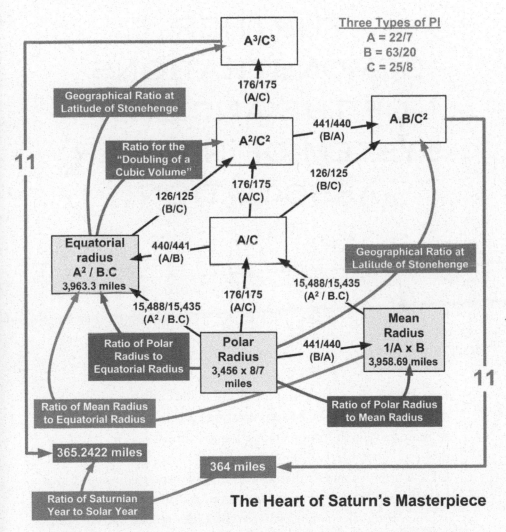

Figure 9.13. The interrelation of three different approximations to pi as found in the microvariations of metrology, summarizing how these relate to the ancient model of the Earth in its three primary radii and to the length of solar year with respect to the otherwise abstract Saturnian year, based upon seven- and thirteen-day weeks.

FURTHER DEMONSTRATIONS OF THE ANCIENT SYSTEM OF NUMBER ASSOCIATION

THE SIGNIFICANCE OF THE MOON

Most briefly it has

1. stabilized the polar axis, promoting stable climatic conditions,
2. eroded the intertidal regions creating crucial habitat for life, and
3. protected Earth from meteoric bombardment.

All this after its genesis through an interplanetary collision, which incidentally led to there being surface metal deposits.*

In other words, the moon got us to where life is today, and at Carnac

---

*For a fuller description of the moon's role in creating life as we know it, see my *Precessional Time and the Evolution of Consciousness,* pages 64–67.

megalithic astronomers discovered the moon and sun were organized in a very specific way, according to their two different years and in an exact geometry of four-squares.

The lunar year is considered to be the 12 whole lunar months contained within the solar year of 365¼ days. The lunar month is the time taken for the moon to complete all of its phases relative to the sun's illumination of the moon, and it is the direct equivalent, for the moon, of the day on Earth. Since the moon orbits the Earth, then it is the Earth's orbit (of 365¼ days) that "moves" the sun, and it is the lunar orbit (of 27⅓ days) that moves the moon to create a synthetic periodicity of both bodies when viewed from Earth, that is, the lunar month (of 29.53 days). Therefore, with respect to the lunar month and lunar year, these cycles are due to the action of three bodies: the Earth's orbit moving the sun along the zodiac, and the moon's orbit, moving around the Earth.

It might be expected, therefore, that the lunar month and year would have no significance and merely be a pattern of illumination, just like those seen throughout the solar system's planets, asteroids, and other moons. Instead, it appears to have some spiritual significance in line with the significance often given it within "traditional" cultures.

TWELVENESS AND THE
SOLAR-LUNAR CALENDAR

The two different years, of the sun and moon, conform to a geometrical structure of only four-squares. This implies that when seen from the Earth, these years have arrived at this condition with some meaning or according to some purpose. Robin Heath (my brother) has been proposing the same geometry as that found at Le Manio since the early 1990s, as it would describe the sun and the moon and can be seen as latent within any 5-12-13 Pythagorean triangle or 5 by 12 rectangle.[1] Hence this geometry of sun and moon could have been behind the 5 by 12 rectangle whose corners are marked by the four Station Stones of Stonehenge. However, we have seen that the megalithic builders in England belonged to a later phase than those at Carnac, where day-inch counting discovered the four-square triangle. This triangle or four-square rectangle was probably common knowledge by the time Stonehenge was built.*

The Quadrilateral at Le Manio is therefore the first and perhaps the only megalithic monument that expressed the four-square relationship clearly. It was not thought that such triangles had been expressions of counting, especially not using the inch to count each day, and so the Quadrilateral has been crucial in revealing this counting technique in the earliest phase of megalithic development. The original megalithic yard emerged and is the length in day-inches of an intercalary month, the difference between three lunar and solar years. The unit was therefore objective in that it harmonized the sun and the moon for the megalithic calendar, showing this monument to be among the most important ever built.

There was a better period for harmonizing the sun and moon, however, 19 years. This generated an even better approximation to the astronomical time difference between solar and lunar years, also producing a more refined megalithic yard, the astronomical megalithic yard (AMY, termed astronomic because of its accuracy). The four-square triangle, when counted in lunar months of one AMY, would, in a single year, release the English foot (the root foot for all later metrology) as its intercalary period per year (rather than 10.875 day-inches).

This transformation operates, from four-squares to 12 lunar months, because it yields the square root of 17 (the four-square diagonal as 4.123. . .), which, when multiplied by three, makes the solar year 12 plus 7/19 lunar months long. It is this that gives the Metonic period closure (as a synodic anniversary) for its cycle of repeated behavior in just 19 years by generating a whole number, 7 lunar months, between 19 lunar and 19 solar years. One way to look at this is to multiply the 4 (squares) by 3 (lunar months) to get 12 (the lunar year) and then multiply the whole diagram again by 19, whereupon 7 will appear as the difference between the diagonal and the base of 4 (squares) times 3 (lunar months per square) times 19 (lunar months), which equals 228 lunar months. This, plus 7, gives the 235 whole lunar months found within the Metonic period, from the geometry alone.

---

*Robin's original view of it, called a Lunation Triangle, has a base of 12 units and a short side of 3 units, hence the four-squares each contain 3 units, just as 12 = 4 times 3.

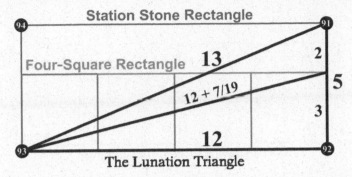

Figure A.I. The solar and lunar years shown in ratio as the Lunation Triangle, here marked within the Station Stone rectangle (of numbered stones) at Stonehenge, which describes a 12 by 5 rectangle. Robin Heath proposed that this gave access to the 12 by 3 rectangle via a division of the 5 side into 3 and 2. Not until 2009 would my theory of day-inch counting and Robin's discovery of this geometry at Carnac converge to show how it had been arrived at a thousand years before Stonehenge.

On a smaller scale, the lunar month might also be considered as being the four along the baseline and the mean solar month as the diagonal, in which case each solar month is slightly longer than a lunar month. This difference is 7/228 lunar months or 0.0307 lunar months.

The new metrology substituted one foot of twelve inches for 10.875 day-inches so that one-twelfth of the solar year, a mean solar month, would be 1 inch greater than the lunar month. Twelve lunar months, counted as megalithic yards, were then 32.625 feet long, and the solar year was 33.625 feet long. The triangle of these measures reduced to normalize its form exposes an N:N+1 ratio where N = 32.625, the megalithic yard found at Carnac. Such a normalization involves the division of both the triangle's longest lengths by their difference in length. This means that the unnormalized lengths, counted in day-inches, were effectively N^2 in length. In the case of the Quadrilateral, three lunar years equaling 1063.1 day-inches divided by the megalithic yard of 32.625 day-inches would have given the length of the AMY, 32.585 inches. It was most fortunate that the three-year anniversary released, through the difference in year lengths, the true length of the megalithic yard (in inches) as the N for this triangle.

The four-square geometry therefore epitomizes the grand unification of 12, 19, and 7 as the organizing principle for the sun and the moon. This geometry could be seen as the very means used in the

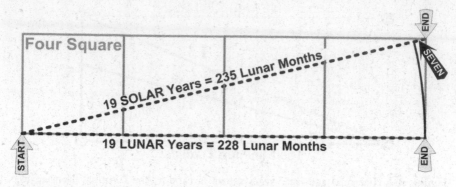

Figure A.2. The four-square geometry taken into the later world of counting months with megalithic yards. This geometry releases, over 19 years, a difference of 7 whole months between lunar and solar years.

development of our planet, in conjunction with its large moon. It is this geometry that causes the orbits of planet and moon to achieve a relatively short synodic period of 19 solar and lunar years, rather than to express the 7 lunar month difference. The megalithic intuited this dual nature of sun and moon, and its integration, as a key to the organization of time and their life on Earth, expressed most directly as the megalithic yard.

The diagonal length of the four-square geometry is 4.123. When the four-square is tripled to a side length of 12, the fractional part of the diagonal (0.123) becomes the 7/19 (0.369) of a month excess of the solar year over the lunar year. In chapters 2 and 3 we saw that the triple-square only approximates the ratio of solar to eclipse years (though quite accurately). In this respect, the triple-square seems more symbolic than exact, especially when one sees that when counting lunar months over one year with megalithic yards, the megalithic astronomers were able to resolve 12/7 feet (the royal cubit) as the difference between the solar and eclipse years using the four-square triangle, as in figure A.3. The royal cubit, found displayed at the heart of Gavrinis's chambered tomb on stone C3, represents the fact that the Saros period of 19 eclipse years is less than the Metonic period by one lunar year of 12 lunar months.

The four-square geometry, which delivered the megalithic yard of three years of day-inch counting, can now deliver the royal cubit, but in the final lunar year of a Metonic period (the last 12 months). One must think "backward" (see figure A.2), since all cyclic repetition is symmetrical relative to its end points.

The excess of 10.875 days in day-inch counting between lunar and solar years was transformed to a foot of 12 inches through the counting of months, using megalithic yards to represent each month. The Saros period of 19 eclipse years was then seen to end before the last 12 lunar months of the Metonic period of 19 solar years. The difference between the Saros and the Metonic was therefore 12 megalithic yards of 19/7 feet, but divided by 19 to expose the difference between a single eclipse year and a single solar year. This cancels the 19 of the AMY (of 19/7 feet) altogether and leaves the difference as 12/7 feet, the royal cubit. The 12 in the royal cubit's numerator actually *is* the twelvefold division of the lunar year and the 7 of its denominator is the sevenfold division of the English foot (as in stone C3): the new intercalary distance between the lunar year and solar year when counted in AMY per lunar month.

The above process means that by using the megalithic yard to count, astronomers of the megalithic could generate the difference between the eclipse year and solar year without using the less-accurate triple-square geometry. Because the result was the 12/7 foot royal cubit, this must

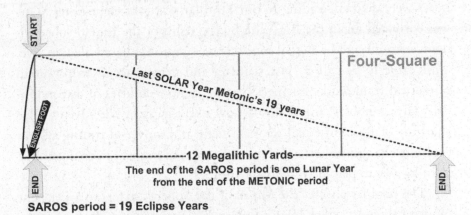

SAROS period = 19 Eclipse Years

Figure A.3. The last solar year of the 19-year Metonic period starts a countdown in which, first, a similar eclipse would repeat ten days into the year (the end of the Saros period), and 12 lunar months after that, the Metonic period returns the sun, moon, and sky to the same relative configuration. It is this last lunar year (of the 19 solar years), counted in one AMY per month, that generates the royal cubit of 12/7 feet, as the excess of the solar year over the eclipse year.

have defined a precedent for the move to making a metrological system dependent upon the English foot, which then stood as the root, or unity, for the sophisticated rationics found in ancient metrology. The unquestionable sacredness of the royal cubit in the ancient Near East must also have had its origin in the 19 eclipse years of the Saros, the twelveness of the lunar year, and the 7 extra lunar months between solar and lunar years in the 19-year Metonic period. This cannot be spoken against factually and must have emerged from the four-square geometry, which plays host to exactly these special characteristics of Earth astronomy.

WHO MADE IT SO?

The esoteric number science of the ancient Near East came to be musical harmonics rather than astronomical periods. Having noted certain musical realities present in astronomical time periods, for example between Jupiter, Saturn, and the lunar month, one can note that ancient musical harmony was primarily founded upon such ratios, which only employed the first three prime numbers: 2, 3, and 5. The first two primes gave rise to Pentatonic tuning but got a bit weird when seven notes were defined, leading to our familiar *octave* of eight notes. This was corrected for in the Rig Veda by the rishis of the Indian subcontinent, who invented a modal music involving tones based also on five. This came to be called "just" tuning and it gave access to powerful numerical frameworks for investigating the possibilities of astronomically large numbers. All of this is now swept away by the advent, since the time of Bach, of equal temperament as a universal tuning system enabling changes of key and standardization of musical instruments, so as to be able to play alongside one another.

The popular phrase "the Music of the Spheres" refers to an ancient mode of thought called *Musica universalis*. From Wikipedia:

Musica universalis (lit. universal music) or Harmony of the Spheres is an ancient philosophical concept that regards proportions in the movements of celestial bodies—the Sun, Moon, and planets—as a form of *musica* (the Medieval Latin name for music). This "music" is not usually thought to be literally audible, but a harmonic and/or mathematical and/or religious concept.[2]

In modern times this proportionality in the movements of celestial bodies has fallen into disrepute as not demonstrable, but the four-square rectangle shows how equal temperament may be found "in the heavens" and the megalithic scientists, for whom this was demonstrable, must be the original source for this doctrine.

Equal temperament divides the world of the octave into twelve equal semitone ratios of 1.05946,* this ratio being the 12th root of two, this often being shown on a circle having 12 equal divisions, like the hours on a clock. We have seen that the lunar year is divided into 12 lunar months and that the solar year can be seen as dividing by 12 into mean solar months, just a little bit larger than the lunar month. Therefore, years seem to be divided into 12 time periods just as, in equal temperament, the octave is divided into 12 semitones. The unique character of four-square geometry enables the ratio of an equal tempered semitone to be generated, relative to the solar year.

The simplest demonstration of this equal-tempered ratio, between the lunar and solar months, is over the three solar year day-inch count, monumentalized within Carnac's Quadrilateral. The difference between two lunar years and three solar years is 387 days, just 1.05946 of the solar year. Hence this constitutes an equal-tempered semitone enlargement of the solar year, corresponding to the stones of the Quadrilateral's southern kerb, abutted stones numbered 24 and 25 (see figure 3.5 on p. 48) from the Sun Gate as stone 1. One might consider this as a mere quirk but the origins of this result lie in the three ratios between the lunar year, the solar year, and the Jupiter synod of 398.88 days.

I first noted that the Jupiter synod was 9/8 longer than the lunar month in my first book, *Matrix of Creation,* and later correlations confirm this as being a part of a cosmic harmonic design involving the

---

*Note how similar 1.05946 is to the length in meters of the hunab at Teotihuacan mentioned in chapter 9. The hunab is indistinguishable from the geographical royal yard, thought of as Egyptian in origin but present at Stonehenge and, as a root cubit of 12/7 feet, at Gavrinis stone C3.

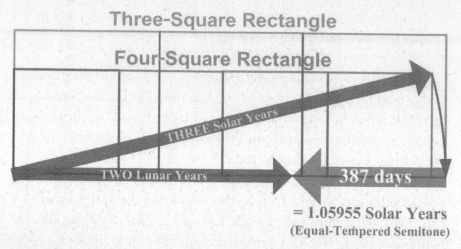

Figure A.4. How the sun and the moon can generate an equal-tempered semitone as the length 387 days, just 1.05946 of the solar year. This length absorbs one lunar year and three excesses of 10.875 day-inches so that two excesses plus the solar year is an equal-tempered semitone.

moon. In hindsight it is obvious that the ratio between the lengths of the lunar year and the Jupiter synod must be the product of two ratios:

1. The ratio between the lengths of the lunar and solar year, which is 1.030689.
2. The ratio between the lengths of the solar year and Jupiter synod, which is 1.092097.

The product of these two ratios is 1.125613, which is one part in 1,836 different from the Pythagorean whole tone of 9/8. What is less expected is that the difference between these same two ratios is 1.05958, one part in ten thousand larger than an equal-tempered semitone. The relationship is even clearer in the diagram shown in figure A.5.

This fact about celestial time can only emerge through the sort of approach taken by the megalithic astronomers. We have seen that numerical coincidences underlie their monuments, demonstrating, even after five to seven thousand years, a means to study astronomy without the arithmetic invented later, in the Near East of the third millennium. It is also clear that ideas such as a demiurge, a harmony of the spheres,

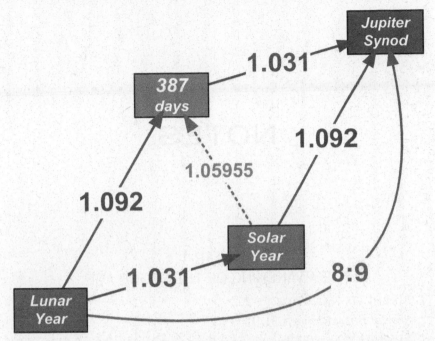

Figure A.5. Diagram of the way in which the two ratios linking the lengths of the lunar year, solar year, and Jupiter synod operate musically to generate a Pythagorean whole tone with their sum and an equal-tempered semitone in their difference.

historical units of measure, and, indeed, our religious ideas came from factual evidence that our world's celestial time environment was a work of number association. Those who later formed religious ideas of a world made up of number association were responding to the facts still true of the world, gleaned during the megalithic era but now all but invisible to the human mind.

NOTES

CHAPTER 1.
THE AWAKENING OF THE STONE AGE

1. Michell, *Ancient Metrology*, 27.
2. *Srimad Bhagavatam*, 4.31.10.
3. Marshack, "The Tai Plaque and Calendrical Notation in the Upper Palaeolithic," 25–61.
4. Ibid.
5. Ibid.
6. Ibid., 27, 35.

CHAPTER 2.
THE TRANSMISSION OF THE SQUARES

1. Schwaller de Lubicz, *The Temple of Man*, ch. 8.
2. de Santillana and von Dechend, *Hamlet's Mill*, chapter on "The Galaxy."
3. Schwaller de Lubicz, *The Temple of Man*, 236.
4. Ibid., 228.

CHAPTER 3.
MEGALITHIC REVELATIONS AT CARNAC

1. Thom, *Megalithic Sites in Britain;* Thom, *Megalithic Lunar Observatories;* and Thom, *Megalithic Remains in Britain and Brittany.*

CHAPTER 4.
THE FRAMEWORK OF CHANGE ON EARTH

1. Helmert, "Ueber die Wahrscheinlichkeit der Potenzsummen der Beobachtungsfehler und über einige damit im Zusam-menhange stehende Fragen."
2. Naydler, *Temple of the Cosmos*, 54–55.
3. de Santillana and von Dechend, *Hamlet's Mill*, 138 and 141.
4. Naydler, *Temple of the Cosmos*, 55.
5. Ibid., 56.
6. Ibid.
7. Ibid., 57.
8. Ibid., 58.
9. Heath, *Precessional Time and the Evolution of Consciousness*, 64.

CHAPTER 5.
LOOKING SOUTH TO MEASURE THE EARTH

1. Ifrah, *The Universal History of Numbers*.
2. Neal, *All Done with Mirrors*.
3. Ibid.
4. From appendix of Thompkins, *Secrets of the Great Pyramid*.
5. de Santillana and von Dechend, *Hamlet's Mill*, 264.
6. John Dowson, *A Classical Dictionary of Hindu Mythology and Religion*.
7. Rig Veda, 5.85.5.
8. de Santillana and von Dechend, *Hamlet's Mill*, 133.
9. Ibid., 134.
10. Proclus' commentary on Plato's *Cratylus*, fr. 155, translated by O. Kern, cited from de Santillana and von Dechend, *Hamlet's Mill*, 134.
11. The Book of Enoch, chapter LXI, verses 1, 2, and 5.

CHAPTER 6.
THE MEGALITHIC MODEL OF THE EARTH

1. Hoyle, *On Stonehenge;* Heath, *Sun, Moon and Stonehenge*.
2. Neal, *All Done with Mirrors*.
3. Berriman, *Historical Metrology*.
4. "Henge," http://en.wikipedia.org/wiki/Henge (accessed February 10, 2013).

5. Neal, *All Done with Mirrors,* 114.

6. Guilbeau, "The History of the Solution of the Cubic Equation."

7. Neal, *All Done with Mirrors,* 72.

8. Ibid., 93.

CHAPTER 7.
FROM EGYPT TO JESUS

1. Gordon and Rendsburg, *The Bible and the Ancient Near East,* 321.

2. Michell, *The Temple at Jerusalem,* addendum.

3. Michell, *Dimensions of Paradise,* 108.

4. "Aion (deity)," http://en.wikipedia.org/wiki/Aion_%28deity%29 (accessed January 10, 2014); "Phanes (mythology)," http://en.wikipedia.org/wiki/Phanes_%28mythology%29 (accessed October 2013).

5. Ulansey, *The Origins of the Mithraic Mysteries.*

6. Steiner, *Real Presences,* 4.

7. McClain, *Myth of Invariance.*

8. Ibid., 143.

9. Bremer, *Plato's Ion: Philosophy as Performance.*

CHAPTER 8.
DESIGNER PLANET

1. Michell, *Dimensions of Paradise,* 34.

2. Ibid., 31.

3. Atkinson, *Stonehenge.*

4. Heath, *Sun, Moon and Stonehenge,* 167–76.

5. Neal, *All Done with Mirrors,* 146–9.

6. Ibid., 139.

7. Helmert, "Ueber die Wahrscheinlichkeit der Potenzsummen der Beobachtungsfehler."

8. Cook, *The Pyramids of Giza.*

9. Heath, "Prehistoric Surveying." Updated PDF available at skyandlandscape.com.

10. Heath, *Sun, Moon and Stonehenge.*

CHAPTER 9.
THE TIME FACTORING OF THE EARTH

1. Gordon, "Channels of Transmission," in *Beyond the Bible.*
2. Aveni, Hartung, and Buckingham, "The Pecked Cross Symbol in Ancient Mesoamerica"; Aveni, *Skywatchers of Ancient Mexico,* appendix A.
3. Rogers, "Origins of the ancient constellations," 9–28.
4. Sullivan, *The Secret of the Incas,* 152.
5. Richards, *Mapping Time,* 390.
6. de Santillana and von Dechend, *Hamlet's Mill,* 129.
7. Ibid., 134.
8. Malhotra, Holman, and Ito, "Chaos and stability of the solar system."
9. de Santillana and von Dechend, *Hamlet's Mill,* 135.

APPENDIX.
FURTHER DEMONSTRATIONS
OF THE ANCIENT SYSTEM OF
NUMBER ASSOCIATION

1. Heath, *Sun, Moon, Man, Woman,* 194; expanded within Heath, *A Key to Stonehenge,* 37.
2. "Musica universalis," http://en.wikipedia.org/wiki/Musica_universalis (accessed October 2013).

GLOSSARY

3-4-5 triangle: The "first" Pythagorean triangle, this triangle is special because the side lengths are successive small primes and define the solstial extremes of the sun at Carnac.

alignments: Name special to megalithic Carnac used, primarily, for three successive groups of parallel rows of stones starting above Carnac called Le Menec, Kermario, and Kerlescan.

astronomical megalithic yard (AMY): A unit of 19/7 feet that reflects the 7-lunar month difference between the Saros and Metonic periods. Actually 19.008/7 feet or, as a rational fraction, 18/7 feet of 1.056 feet.

Aubrey Circle: The circle of 56 Aubrey holes at Stonehenge inside the bank and ditch, far older than the famous stone circle at Stonehenge.

azimuth: The standard angular measure on the horizon, measured clockwise from north, which marks zero degrees.

boustrophedon: A system of folding a time line to fit a smaller area, seen in the serpentine path of recorded time on the Tai Plaque.

Canevas: A technique of geometrical knowledge expressed within multiple square grid, noted and named by Schwaller de Lubicz.

canonical: A type of root measure varied by 126/125 (1.008), after John Michell.

Carnac: A small town in southern Brittany, France, synonymous with northwest Europe's largest megalithic complex from the fifth millennium BCE.

circumpolar observatory: A system for tracking the motion of circumpolar stars around a normally starless north pole using a foresight (stationary pole or hilltop) to mark multiple backsights and record the motion of the stars.

cromlech: A French word (from Breton) for round kerb monument, rather than a stone circle, seen at Carnac at the termination of the long parallel stone rows called alignments.

cubit: A unit of 3/2 feet of any sort, such as 12/7 of royal feet.

day-inch counting: The fifth millennium BCE practice of using an inch to represent each day.

diffusionism: A popular doctrine that claimed all civilizing information had diffused from the civilizations of the Fertile Crescent. Now seriously questioned, in particular by Colin Renfrew in his *Before Civilisation,* after carbon dating techniques and in particular their required corrections, gave cultural dates that made transmission from the east to the west unlikely if not impossible.

dolmen: A chamber made of vertical megaliths upon which a roof slab is balanced.

eclipse year: The time taken for the sun to again sit on the same lunar node, when it is then very likely to take part in an eclipse: 346.62 days.

English foot: The standard ancient foot, representing unity, of which all other measures are rational fractions or microvariations.

Er Grah: The foundation hole in Locmariaquer where large menhirs were erected and beside which the largest standing stone in Europe lies broken—Grand Menhir Brisé.

geodetic: Relating to the numerical definition of the shape of the Earth.

geodetic menhir: A standing stone used to mark distances between places or used for the establishment of latitude and, in concert, longitude.

gnomic pole: A vertical pole used to track solar time, establish true north and the north pole, and find geocentric latitude.

gnomon: An ancient Greek word meaning "knowledge," "indicator," "one who discerns," or "that which reveals." The root of "gnomic pole."

Grand Menhir Brisé: The largest standing (and sculpted) stone in Europe, now broken and no longer standing, parts of which form the roofs of nearby dolmen Table des Marchands and the chamber of Gavrinis Cairn.

hunab: A unit of length of about 1.05945 meters found at Teotihuacan by H. Harleston Jr. probably a geographic royal yard though also the twelfth root of 2.

Hyperboreans: The name given by the ancient Egyptians and classical Greeks to "men from the north."

inverse Roman foot: Perhaps the first geodetic foot whose microvariations are also rational, 25/24 feet.

Jerusalem cubit: The Egyptian royal cubit enlarged by the canonical microvariation, 1.728 feet.

Jupiter synod: The 398.88 days between retrograde loops of Jupiter, equal to 57 seven-day weeks.

kerb: A row of stones, usually small and flat, found abutted to a megalithic monument.

Kerlescan alignments: The stone rows immediately east of Kermario.

Kermario alignments: The stone rows immediately east of Le Menec's alignments.

Le Manio: A small hill north of the Kerlescan alignments upon which Menhir Geant, visible from distant sites, and the Quadrilateral were constructed in the late fifth millennium BCE.

Le Menec: A small hamlet built within the western cromlech of the Le Menec alignments.

lunar month: The time taken for the sun's illumination of the moon to complete its set of phases seen from the Earth.

lunar nodes: The two points at which the moon, in its orbit, crosses the sun's path or ecliptic; eclipses occur at those points.

lunar year: The twelve whole lunar months that fit within a solar year, which is 12 7/19 lunar months long.

Lundy Island: A hill inundated thousands of years ago by rising sea levels to form an island in the Bristol Channel exactly west of Stonehenge by 240,000 AMY and exactly south of the source site for the Stonehenge bluestones.

Mane Lud: The northern tumulus terminating the Locmariaquer complex.

Mane Ruthual: A large multi-chambered dolmen south of Er Grah in Locmariaquer.

megalithic inch: One-fortieth of a megalithic yard, about 0.815 of an inch, a division applied to similar lengths, such as 33 inches.

megalithic yard: At Le Manio 32.625 day-inches. See also "astronomical megalithic yard (AMY)."

menhir: A Breton word for a standing stone.

Menhir Geant: The large menhir of the Le Manio complex.

Metonic period: A cycle of 19 years, in which any configuration of sun, moon, and stars will repeat.

metrology: The use of replicable lengths to count and store numbers and make measurements.

multiple squares: A technique used at Carnac enabling the layout of monuments to easily align to astronomical horizon events or time periods and where the diagonal angles relate in combination to the angle of another multiple square.

Octon eclipse period: A first but lesser eclipse period to the Saros, of 47 lunar months approximating four eclipse years, and related to the Metonic period.

Ouroboros: An ancient symbol of reentrancy in which, like an orbit, a serpent is shown eating its tail.

Precession: The tendency of any rotating or orbiting body, when perturbed, to wobble cyclically in retrograde, through a force at right angles to its polar axis.

Precession of the Equinoxes: The effect on the Earth of precession of its polar axis, seen in the retrograde movement of the two crossings of ecliptic and celestial equator, anciently divided into Ages named after the zodiacal constellations at the spring (or vernal) equinox.

Quadrilateral: A kerb structure at Le Manio near Carnac.

retrograde: Motion opposite to that of the sun and planets, toward the celestial west.

royal foot, royal cubit, and royal yard: The royal foot is 8/7 English feet, the royal cubit is 3/2 of the royal foot (12/7 English feet), and the royal yard is 3 royal feet (24/7 English feet).

Saros eclipse cycle: The 223 lunar month period coinciding with nodal and anomalistic months after 19 eclipse years, guaranteeing a similar eclipse.

Sarsen Circle: A circle of standing sculpted sandstone at Stonehenge with horizontal lintels spanning their tops to form an annular ring.

Saturnian year: 364 days or 52 weeks of 7 days, 13 months of 28 days that related to the Saturn synod.

Saturn synod: The 378 days between two retrograde loops of Saturn, equal to 54 seven-day weeks.

sidereal day: Time for the Earth's rotation relative to the stars, in practice 365/366 of a solar day.

solar day: The time taken for the rotation of the Earth relative to the sun.

solar year: The time taken for the sun to be at the same point on the ecliptic after the Earth's orbit of the sun.

standard canonical Persian foot: A foot of 1.056 feet, proposed as geodetic in dividing the mean Earth circumference.

sublunary sphere: A concept derived from Greek astronomy as the region of the geocentric cosmos from the Earth to the moon, made of the four classical elements: earth, water, air, and fire beyond which the planets and stars are made of aether. After Plato and Aristotle.

Sumerian foot: 12/11 [1.09] feet.

superparticular triangle: A right triangle where the two longest sides have the relationship of N:N+1.

synod: Literally "meeting," a periodic celestial event involving two or more celestial bodies, by their aspect: conjunction, opposition, or elongation.

Table des Marchands: A dolmen at Lochmariaquer near Carnac named after a merchant family.

Teotihuacan: Olmec City in Great Valley of Mexico, with large sacred precinct: two pyramids aligned to Thurban (chapter 9).

Theodolite: Instrument used in surveying consisting of azimuth and altitude graduations aligned to a telescope.

time-factored bones: A technique of marking days as marks on bones.

tryptic: A set of three pictures displayed as center, left, and right, usually religious.

tumulus: An artificial mound, hill, or earthworks usually covering buried elements.

western alignments: The alignments east of Le Menec.

BIBLIOGRAPHY

Atkinson, Richard J. C. *Stonehenge*. Revised edition. Harmondsworth: Penguin Books, 1979.

Aveni, A., H. Hartung, and B. Buckingham. "The Pecked Cross Symbol in Ancient Mesoamerica." *Science* 202, no. 4365 (1978): 267–86.

Aveni, Anthony. *Skywatchers of Ancient Mexico*. Austin, Texas: University of Texas Press, 2001.

Berriman, A. E. *Historical Metrology*. London: Dent, 1953.

Bremer, John. *Plato's Ion: Philosophy as Performance*. North Richland Hills, Tex.: Bibal Press, 2005.

Charles, R. H., trans. *The Book of Enoch*. London: SPCK, 1982.

Cook, Robin. *The Pyramids of Giza*. Glastonbury: Seven Islands, 1992.

Crowhurst, Howard. *Carnac: The Alignments*. Plouharnel: Epistemea, 2011.

———. *Mégalithes*. Plouharnel: HCom Editions, 2007.

Darlisson, Bill. *The Gospel and the Zodiac*. London: Duckworth Overlook, 2007.

Dart, John. *Decoding Mark*. London: Continuum, 2006.

de Santillana, Giorgio, and Hertha von Dechend. *Hamlet's Mill*. Boston: David R. Godine, 1977.

Dowson, John. *A Classical Dictionary of Hindu Mythology and Religion*. New Delhi: Metropolitan Book Co., 1978.

Flammarion, Camille. *The Flammarion Book of Astronomy*. London: Allen & Unwin, 1964.

Gordon, Cyrus H. *Beyond the Bible*. London: Collins, 1962.

Gordon, Cyrus H., and Gary A. Rendsburg. *The Bible and the Ancient Near East*. New York: W. W. Norton, 1997.

Guilbeau, Lucye. "The History of the Solution of the Cubic Equation." *Mathematics News Letter* 5, no. 4 (1930): 8–12.

Hall, Sidney. *Urania's Mirror* (set of 32 constellation cards). London: Samuel Leigh of the Strand, 1825.

Heath, Richard. *Matrix of Creation.* Rochester, Vt.: Inner Traditions, 2002.

———. *Precessional Time and the Evolution of Consciousness.* Rochester, Vt.: Inner Traditions, 2011.

———. *Sacred Number and the Origins of Civilization.* Rochester, Vt.: Inner Traditions, 2007.

Heath, Robin. "Prehistoric Surveying." 2008. Updated PDF available at skyandlandscape.com.

———. *A Key to Stonehenge.* Cardigan, Wales: Bluestone Press, 1993.

———. *Sun, Moon and Stonehenge.* Cardigan, Wales: Bluestone Press, 1998.

———. *Sun, Moon, Man, Woman.* Cardigan: privately printed, 1992.

Heath, Robin, and John Michell. *The Measure of Albion.* Cardigan, Wales: Bluestone Press, 2004.

Helmert, Friedrich Robert. "Ueber die Wahrscheinlichkeit der Potenzsummen der Beobachtungsfehler und über einige damit im Zusam-menhange stehende Fragen." *Zeitschrift für Mathematik und Physik* 21 (1876): S. 102–219.

Hoyle, Fred. *On Stonehenge.* San Francisco: W. H. Freeman, 1977.

Ifrah, Georges. *The Universal History of Numbers.* London: The Harvill Press, 1998.

Ivimy, John. *The Sphinx and the Megaliths.* London: Turnstone, 1974.

MacKie, Euan. *Science and Society in Prehistoric Britain.* London: Paul Elek, 1977.

Malhotra, Renu, Matthew Holman, and Takashi Ito. "Chaos and stability of the solar system." *Proceedings of the National Academy of Sciences* 98, no. 22 (2001).

Marshack, Alexander. "The Tai Plaque and Calendrical Notation in the Upper Palaeolithic." *Cambridge Archaeological Journal* 1 (1991): 25–61.

McClain, Ernest. *Myth of Invariance.* New York: Nicolas-Hays, 1978.

———. *The Pythagorean Plato: Prelude to the song itself.* New York: Nicolas-Hays, 1976.

Michell, John. *Ancient Metrology.* Bristol, U.K.: Pentacle Books, 1981.

———. *Dimensions of Paradise.* Rochester, Vt.: Inner Traditions, 2008.

———. *The Temple at Jerusalem: A Revelation.* Glastonbury: Gothic Image, 2000.

Naydler, Jeremy. *Temple of the Cosmos.* Rochester, Vt.: Inner Traditions, 1996.

Neal, John. *All Done with Mirrors.* London: Secret Academy, 2000.

Needham, Joseph. *Science and Civilisation in China.* Volume 3: *Mathematics and the Sciences of the Heavens and the Earth.* Cambridge, U.K.: Cambridge University Press, 1959.

Richards, E. G. *Mapping Time.* New York: Oxford University Press, 1998.

Rogers, John H. "Origins of the ancient constellations: I. The Mesopotamian traditions." *Journal of the British Astronomical Association* 108 (1998): 9–28.

Schwaller de Lubicz, R. A. *The Temple of Man.* 2 volumes. Translated by Deborah Lawlor and Robert Lawlor. Rochester, Vt.: Inner Traditions, 1997.

Spanuth, Jurgen. *Atlantis of the North.* London: Sidgwick & Jackson, 1979.

Steiner, George. *Real Presences.* London: Faber and Faber, 1989.

Sullivan, William. *The Secret of the Incas.* New York: Crown, 1996.

Thom, Alexander. *Megalithic Lunar Observatories.* London: Oxford University Press, 1971.

———. *Megalithic Remains in Britain and Brittany.* London: Oxford University Press, 1978.

———. *Megalithic Sites in Britain.* London: Oxford University Press, 1967.

Thomkins, Peter. *Secrets of the Great Pyramid.* London: Harper Row, 1971.

Ulansey, David. *The Origins of the Mithraic Mysteries.* New York: Oxford University Press, 1991.

West, John Anthony. *Serpent in the Sky.* London: Wildwood House, 1979.

INDEX

Page numbers followed by "n" indicate notes.
Pages numbers followed by "*f*" indicate figures.